AMBRIDGE

Higher

MATHEMATICS
GCSE for OCR
Homework Book

Nick Asker and Karen Morrison

CAMBRIDGE
UNIVERSITY PRESS

University Printing House, Cambridge CB2 8BS, United Kingdom

One Liberty Plaza, 20th Floor, New York, NY 10006, USA

477 Williamstown Road, Port Melbourne, VIC 3207, Australia

314–321, 3rd Floor, Plot 3, Splendor Forum, Jasola District Centre, New Delhi – 110025, India

79 Anson Road, #06–04/06, Singapore 079906

Cambridge University Press is part of the University of Cambridge.

It furthers the University's mission by disseminating knowledge in the pursuit of education, learning and research at the highest international levels of excellence.

www.cambridge.org
Information on this title: www.cambridge.org/ukschools/9781107496927 (Paperback)

First published 2015

20 19 18 17 16 15 14 13 12 11 10 9 8 7 6 5 4

Printed in Great Britain by Ashford Colour Press Ltd.

A catalogue record for this publication is available from the British Library

ISBN 978-1-107-49692-7 Paperback

Additional resources for this publication at www.cambridge.org/ukschools

Cover image © 2013 Fabian Oefner www.fabianoefner.com

....

NOTICE TO TEACHERS IN THE UK

....

This resource is endorsed by OCR for use with specification J560 GCSE Mathematics.

In order to gain OCR endorsement this resource has undergone an independent quality check. OCR has not paid for the production of this resource, nor does OCR receive any royalties from its sale. For more information about the endorsement process please visit the OCR website **www.ocr.org.uk**

The questions in this book are not necessarily indicative of the question wording and style that will appear in OCR GCSE examinations.

Contents

Introduction

This book has been written by experienced teachers to help you practise applying the skills and knowledge you will learn during your GCSE course.

Each chapter is divided into sections, which cover individual topics. A section contains one or more homework exercises, containing a range of questions so you can apply your knowledge of the topic. At the end of each chapter, the Chapter review contains a mixture of questions covering all of the topics in the chapter.

Look out for the following features throughout the book:

 This means you might need a calculator to work through a question.

 This means you should work through a question without using a calculator. If this is not present, you can use a calculator if you need to.

Tip

Tip boxes provide helpful hints.

The Homework Book chapters and sections match those of the *GCSE Mathematics for OCR Higher Student Book*, so you can easily use the two books alongside each other. However, you can also use the Homework Book without the Student Book.

You can check your answers using the free answer booklet available at **www.cambridge.org/ukschools/gcsemaths-homeworkanswers**.

1 Basic calculation skills

Section 1: Basic calculations
HOMEWORK 1A

Solve these problems using written methods.
Set out your solutions clearly to show the methods you chose.

1. What do you need to add to each number to make 7?
   a 8   **b** 13   **c** ⁻2

2. What would you subtract from each number to get a result of ⁻12?
   a 8   **b** 6   **c** ⁻4

Tip
Learn the rules for calculations with negative numbers.

3. ⁻7 is multiplied by another number to get each result. Work out what the other number is in each case.
   a 14   **b** ⁻98   **c** ⁻49   **d** 343   **e** ⁻7

4. By what would you divide ⁻96 to get the following results?
   a 8   **b** ⁻8   **c** 2   **d** $-\frac{1}{2}$ **e** ⁻256

5. Here is a set of integers {⁻7, ⁻5, ⁻1, 2, 7, 11}.
   a Find two numbers with a difference of 7.
   b Find two numbers with a product of ⁻7.
   c Find three different numbers with a sum of 4.
   d Find two numbers which, when divided will give an answer of ⁻1.

6. Two more than ⁻8 is added to the product of 8 and five less than 3.
   What is the result?

HOMEWORK 1B

1. How many 12 litre containers can be completely filled from a tanker containing 783 litres?

2. A train is travelling at a constant 64 mph.
   a How far does it travel in $1\frac{1}{2}$ hours?

   b How long does it take to travel 336 miles?

Tip
64 mph means the train travels 64 miles each hour.

3. A train starts a journey with 576 people on board.
   At the first station 23 people get on, 14 get off.
   At the second station 76 people get off and no one gets on.
   At the third station a further 45 people get on.
   How many people are on the train after the third station?

4. Henry goes shopping with £125. He spends £26 on a DVD, £38.19 on a jumper and gets three books at £2.85 each.
   How much does he have left when he returns home?

5. If six cups of coffee cost £11.70 and three cups of tea cost £4.23, how much would four cups of coffee and five cups of tea cost?

6. Find the difference between the product of 17 and 51 and the sum of 156 and 652.

Section 2: Order of operations
HOMEWORK 1C

1. The temperature one day in Aberdeen is 3°C. Overnight the temperature drops by 11°C. What is the temperature overnight?

2. Calculate.
   a 13 − 4 + 8   **b** ⁻4 − 3 − 7
   c ⁻5 + 9 − 6   **d** ⁻8 − (⁻5) + 3
   e ⁻27 + (⁻12) − 18

3. Insert brackets into each calculation to make it true.
   a 4 × 5 + 7 = 48   **b** 35 − 20 × 8 = 120
   c 48 ÷ 4 × 3 − 4 = ⁻12

Tip

Remember, brackets change the order of operations.

4 Each ☐ represents an operation.
Fill in the missing operations to make these statements true.

a $12 \,\square\, (36 \,\square\, 32) = 3$

b $95 \,\square\, (13 \,\square\, 8) = 19$

5 Work out without using a calculator:

a $\dfrac{10 \times \sqrt{25}}{3^2 + 4^2}$

b $\dfrac{6^2 \times \sqrt{4}}{2^2 \times 3^2}$

c $\dfrac{\sqrt{4} + 5^2}{3^2 \times \sqrt{81}}$

d $\dfrac{6^2 + 8^2}{12^2 - (4^2 \times \sqrt{9}) + 2^2}$

Section 3: Inverse operations
HOMEWORK 1D

Tip

Remember, the inverse is the 'opposite' operation that reverses the effect of an operation.

1 Simplify.

a $6 \times 11 + 4$

b $6 \times (11 - 2)$

c $5 + 11 \times 2$

d $(3 + 12) \times 4$

e $25 + 6 \times 3$

f $8 \times 3 \div (4 + 2)$

g $(14 + 7) \div 3$

h $43 + 2 \times 8 + 6$

i $24 \div 4 \times (8 - 5)$

j $16 - \dfrac{8}{2} + 5$

2 Use the numbers listed to make each number sentence true.

a $\square - \square \div \square = \square$ 1, 18, 6, 4

b $\square - \square \div \square = \square$ 8, 7, 3, 2

c $\square \div (\square - \square) - \square = \square$ 2, 3, 4, 7, 15

Tip

Learn the rules about order of operations.

3 Use inverse operations to check the results of each calculation.

Correct those that are incorrect.

a $50 \times 5 - 8 = 227$

b $16 + 5 \times 8 - 12 = 50$

c $(28 + 53) \times 4 = 264$

d $(432 - 148) \div 4 = 71$

4 Use inverse operations to find the missing values in each of these calculations.

a $\square + 564 = 729$

b $\square + 389 = 786$

c $\square - 293 = 146$

d $132 \times \square = {}^-3564$

e $^-8 \times \square = 392$

f $\square \div 30 = 4800$

5 The formula for finding the area of a triangle is $A = \dfrac{bh}{2}$.

Find the height of a triangle if it has an area of 72 cm^2 and a base length of 8 cm.

Chapter 1 review

1 Bonita and Kim travel for $3\frac{1}{2}$ hours at 48 km per hour.
They then travel a further 53 km.
What is the total distance they have travelled?

2 On a page of a newspaper there are 8 columns of text.
Each row contains a maximum of 38 characters.
Each column has a total of 168 rows.

a What is the maximum number of characters that can appear on a page?

b The average word length is six characters, and each word needs a space after it.
Estimate the number of words that can fit on a page.

3 A theatre has seats for 2925 people. How many rows of 75 is this?

4 Two numbers have a sum of $^-12$ and a product of $^-28$. What are the numbers?

5 Jadheja's bank account was overdrawn.
She deposited £750 and this brought her balance to £486.
By how much was her account overdrawn to start with?

6 We use the formula $F = 2C + 32$ to convert temperatures from Celsius to Fahrenheit.

Without using a calculator, find the temperature in degrees Celsius when it is:

a 78°F

b 120°F

2 Whole number theory

Section 1: Review of number properties

HOMEWORK 2A

1 Look at the numbers in this list:

4 15 8 25 7 16 12 9
6 3 36 96 27 3 1

Write down the numbers in the list that are:
a odd
b even
c prime
d square
e cube
f factors of 12
g multiples of 4
h common factors of 24 and 36
i common multiples of 3 and 4.

> **Tip**
>
> Check you know the words for different types of numbers.

2 Write down:
a the next four odd numbers after 313.
b the first four consecutive even numbers after 596.
c the square numbers between 40 and 100 inclusive.
d the factors of 43.
e four prime numbers between 30 and 50.
f the first five cube numbers.
g the first five multiples of 7.
h the factors of 48.

3 Say whether the results will be odd or even.
a The product of two odd numbers.
b The sum of two odd numbers.
c The difference between two odd numbers.
d The square of an even number.
e The product of an odd and an even number.
f The cube of an even number.

HOMEWORK 2B

1 Write these sets of numbers in order from smallest to biggest.
a 476 736 458 634 453 4002
b 1707 1770 1708 1870 1807
c 345 543 453 354 534 435

2 What is the value of the 6 in each of these numbers?
a 46   b 673   c 265
d 16 877   e 64 475   f 1 654 782
g 6 035 784

3 What is the biggest and smallest number you can make with each set of digits?
Use each digit only once in each number.
a 3, 0 and 7   b 6, 5, 1 and 9
c 2, 3, 5, 0, 6 and 7
d What is the difference between the biggest and smallest numbers in each question?

Section 2: Prime numbers and prime factors

HOMEWORK 2C

1 Identify the prime numbers in each set.
a 10, 11, 12 ,13 ,14 , 15, 16, 17, 18, 19, 20
b 100, 101, 102, 103, 104, 105, 106, 107, 108, 109, 110

> **Tip**
>
> Remember each number has a unique set of prime factors.

2 Express each of the following numbers as a product of their prime factors.

Use the method you prefer. Write your final answers using powers.
a 48   b 75   c 81   d 315
e 560   f 2310 g 735   h 1430
i 32   j 635   k 864

3 A number is expressed as $13 \times 23 \times 7$.
What is the number?

4
  a   What are the factors of 7120?
  b   What are the factors of 2279?
  c   What do you notice about the factors of 2279?
  d   Did it take longer to find the factors of 2279 or 7120?

5
  a   Calculate 31×67.
  b   What are the factors of 2077?

Section 3: Multiples and factors
HOMEWORK 2D

1 Find the LCM of the given numbers.
  a   12 and 16
  b   15 and 20
  c   2 and 20
  d   24 and 30
  e   3, 4 and 6
  f   5, 7 and 10

2 Find the HCF of the given numbers.
  a   18 and 24
  b   36 and 48
  c   27 and 45
  d   14 and 35
  e   21 and 49
  f   36 and 72

3 Find the LCM and the HCF of the following numbers by using prime factors.
  a   28 and 98
  b   75 and 20
  c   144 and 24
  d   54 and 12
  e   214 and 78

4 Amjad has two long pieces of timber.
One piece is 64 metres, the other is 80 metres.
He wants to cut the long pieces of timber into shorter pieces of equal length.
What is the longest he can make each piece?

> 💡 **Tip**
>
> Think carefully – is it the HCF or the LCM you need to find?

5 Two desert flowers have a life cycle of 11 and 15 years respectively, when they are in bloom. How many years are there between the occasions when they bloom simultaneously?

6 Rochelle has 20 bananas and 55 toffees to share among the pupils in her class.
She is able to give each student an equal number of bananas and an equal number of toffees.
What is the largest possible number of students in her class?

7 Mr Singh wants to tile a rectangular patio with dimensions 5.4 m $\times$ 9.6 m with a whole number of identical square tiles.
Mrs Singh wants the tiles to be as large as possible.
  a   What is the largest possible square tile that will fit exactly along both dimensions? Show your working.
  b   Find the area of the largest possible tile in cm^2. Show your working.
  c   How many tiles will Mr Singh need to tile the patio? Show your working.

Chapter 2 review

1 Is 243 a prime number? Explain how you worked out your answer.

2 Find the HCF and the LCM of 18 and 45 by listing the factors and multiples.

3 Express 675 as a product of prime factors, giving your final answer in power notation.

4 Determine the HCF and LCM of the following by prime factorisation.
  a   64 and 104   b   54 and 80
result is divided by 3.

5 Mimi starts her training for a triathlon on Monday May 2nd. She swims and cycles on this day.
She decides to swim every third day, run on Wednesdays and Saturdays and cycle every fourth day.
On which date in May will she swim, run and cycle on the same day?

3 Algebraic expressions

Section 1: Algebraic notation

HOMEWORK 3A

1. Write an expression for the following statements using the conventions for algebra.

> **Tip**
>
> Remember – letters represent numbers.

   a A number x is multiplied by 4 and has 3 added to it.
   b A number x is multiplied by 2 and added to the number y that has been multiplied by 5.
   c A number x is squared and 7 is subtracted from this, and it is all multiplied by 3.
   d A number x is cubed and added to a number y squared and this is all divided by 2.
   e A number x has 2 subtracted from it and the result is divided by 3.

2. Match each statement to the correct algebraic expression.

A number x is multiplied by 2 and has 7 added to it. The result is divided by 3.	$3x - 7$
A number x is squared, then multiplied by 3, and added to a number y multiplied by 7.	$x^2 + 3x$
A number x is multiplied by 3 and has 7 taken from the result.	$3x^2 + 7y$
A number x is added to a number y, and the result is multiplied by 3.	$\dfrac{(2x + 7)}{3}$
A number x is squared, and the same number is multiplied by 3. These are then added together.	$3(x + y)$

3. Simplify these expressions.
   a $5 \times 2x$
   b $3a \times 2$
   c $x \times (^-5)$
   d $3x \times 6y$
   e $3a \times 5b$
   f $^-3p \times 3q$
   g $16x \div 4$
   h $25y \div 5$
   i $32a^2 \div 4$
   j $6 \times 15p \div 20$
   k $27x \div (3 \times 3)$
   l $24y \div (4 \times 2)$

4. Write an expression to represent the area of each of these rectangles.

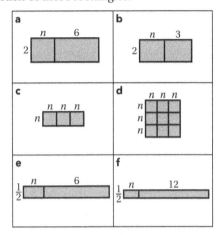

5. Simplify.
   a $y^{-2} \times y^6$
   b $2a^5 \times 3a^4$
   c $a^3 \div a^6$
   d $\dfrac{18b^3}{6b^2}$
   e $\dfrac{24p^8}{6p^9}$
   f $(3a^6)^2$
   g $3x^3y^2 \times 6xy^5 \div 3x^4y^5$
   h $10x^0$

HOMEWORK 3B

1. Given that $p = 4$ and $q = 5$, find the value of these expressions.
   a $4p + 2q$
   b $2p - q$
   c $3pq$
   d $2q + 3p$

2. Find the value of each expression when $e = {^-2}$ and $f = 6$.
   a $3e + 2f$
   b ^-4ef
   c $\dfrac{100}{e}$
   d $6 - 5ef$

3. Given that $x = 4$ and $y = 3$, evaluate the following expressions.
   a $3(2x + 3y)$
   b $^-2(x + 2y)$
   c $3y(2x - y)$
   d $5(10 - 2y)$

5

Section 2: Simplifying expressions
HOMEWORK 3C

1. Which of the following pairs are like terms? Collect where possible.
   - **a** $10x$ and $4a$
   - **b** $8b$ and ^-3b
   - **c** $9m$ and $6n$
   - **d** ^-8xy and ^-5y
   - **e** $6pq$ and ^-3p
   - **f** $10x^2$ and $5x^2$
   - **g** $7x^2$ and $^-7x^2$
   - **h** $6x^2$ and ^-2x
   - **i** $3a^2bc$ and $4a^2bc^2$

Tip

You cannot add $3a$ to $4b$ unless you know the values of a and b!

2. Write these expressions in their simplest form by collecting up like terms.
   - **a** $3a + 6b - 7a + 4b$
   - **b** $6a + 9b - 5a - 8b$
   - **c** $4ab + 5b^2 + 7ab - 7b^2$
   - **d** $4m^2 - mn^2 + mn^2 + 6mn$
   - **e** $8cd^3 - 24cd^3 + 5cd^3$
   - **f** $4st^2 - 4s^2t + 7s^2t + 5st^2$

3. Copy and complete.
   - **a** $4a + \square = 10a$
   - **b** $7b - \square = 6b$
   - **c** $12mn + \square = 15mn$
   - **d** $17pq + \square = 8pq$
   - **e** $9x^2 - \square = 12x^2$
   - **f** $8m^2 - \square = ^-m^2$
   - **g** $6ab - \square = ^-2ab$

4. Copy and complete.
   - **a** $6a \times \square = 18a$
   - **b** $7b \times \square = 14b$
   - **c** $4a \times \square = 12ab$
   - **d** $7m \times \square = 28mn$
   - **e** $^-4b \times \square = 12b^2$
   - **f** $6m \times \square = 12m^2n$

5. Cancel to lowest terms to simplify.
   - **a** $\dfrac{6x}{2}$
   - **b** $\dfrac{4a}{12}$
   - **c** $\dfrac{^-16m}{24}$
   - **d** $\dfrac{14x^2}{21}$
   - **e** $\dfrac{9ab}{a}$
   - **f** $\dfrac{4xy}{12xy}$

Section 3: Multiplying out brackets
HOMEWORK 3D

1. Expand the brackets and collect any like terms to simplify the following:
   - **a** $3(a + 4) - 9$
   - **b** $4(a - 3) + 2$
   - **c** $6(b + 4) - 10$
   - **d** $4(e - 6) + 17$

   - **e** $3(x - 7) - 4$
   - **f** $3a(2a + 5) + 8a$
   - **g** $3b(4b - 7) - 6b$
   - **h** $3a(4a + 7) + 5a^2$
   - **i** $5b(4b - 5) - 9b^2$

Tip

Multiply everything inside the bracket by the number outside.

2. Expand and collect like terms for each expression.
   - **a** $3(x + 2) + 4(x + 5)$
   - **b** $3(4a - 1) + 4(3a - 2)$
   - **c** $4(c + 6) - 3(c + 7)$
   - **d** $4(a - 3) - 3(a + 4)$
   - **e** $x(x - 5) + 2(x - 7)$
   - **f** $5q(q + 3) - 5(q + 2)$
   - **g** $2y(y + 5) - y(2y + 3)$
   - **h** $2x(x - 5) + x(x - 3)$

3. In the diagram below, each box is the sum of the two boxes below it. Find the missing expressions for the empty boxes.

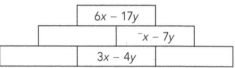

Section 4: Factorising expressions
HOMEWORK 3E

1. Which of the following are expressions, identities or equations? Correctly use the symbol $\equiv$ for any that you think are identities.
   - **a** $3(a + b) = 3a + 3b$
   - **b** $x + y = y + x$
   - **c** $xy^2 + x^2$
   - **d** $3a + 5$
   - **e** $4a + 3b = 22$
   - **f** $x(x + 1) = x^2 + x$

2. Write an expression for the sum of three consecutive even numbers.
   Prove that the sum is a multiple of three.

Section 5: Using algebra to solve problems
HOMEWORK 3F

1. Copy and complete this magic square, filling in the missing expressions.

   a

$3n + 8$	$3n - 13$	$3n + 2$
	$3n - 1$	

   b Write an expression for the magic number.

2 A large rectangle contains a smaller rectangle.

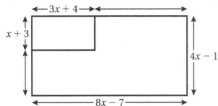

a Write an expression for each of the missing lengths.

b Write an expression for the perimeter of the small rectangle.

c Write an expression for the perimeter of the large rectangle.

d Write an expression for the perimeter of the compound shape formed by removing the small rectangle from the large rectangle.

3 In the diagram below, each box is the sum of the two boxes below it. Find the missing expressions for the empty boxes.

$4x^2 + x - 12y$	
$4x - 8y$	
	$^-3x - 5y$

4 Which of the following are always, sometimes or never true?
If the answer is sometimes, state when it is true.

	Always true	Sometimes true when...	Never true
$x + 4 = 7$			
$3x - 4 = 4 - 3x$			
$2x - 4 = 2y - 4$			
$3(n - 4) = 3n - 12$			
$x^2 + 3x + 4 = 4 + x(x + 3)$			

5 Follow the instructions and record your answer at each stage.
1. Think of a number.
2. Multiply it by 3.
3. Add 1 to the result.
4. Multiply the answer by 2.
5. Add 10 to the result.
6. Divide the answer by 6.
7. Take away the number you first thought of.
8. Your answer is 2.
Prove that the result will always be 2

6 A sequence begins with two numbers, a and b. The next number in the sequence is always found by adding the previous term to double the term before the previous term, e.g. the third term is $2a + b$.

a Prove that the 7th number in the sequence is always $22a + 21b$.

b If a and b are $^-3$ and 2, what would the value of the 7th number be?

Chapter 3 review

1 The expression $6(x + 3) - 4(x - 3)$ simplifies to $a(x + b)$.
Work out the values of a and b.

2 Simplify the following expressions fully by collecting like terms where possible.

a $4(x + 3) + 8x - 5x + 12 + 7x$

b $4(x - 3) + 3(x - 4)$

c $2x(x - 3) + 3x - x^2$

d $\dfrac{21x^3}{3x} + \dfrac{6x^3}{x}$

e $2x(3x + 8) - 8x$

f $4a(4a - 3) - 4b$

g $4a(5a - 6) + 3a^2$

h $3x(4x - 5) - 5x^2$

3 Which of these equations are identities? Correctly use the symbol $\equiv$ for any that you think are identities.

a $6x + 4 = 3x + 2$ b $5xy + 3 = 3 + 5xy$

c $x^2 = 2x$ d $x(y + 7) = xy + 7x$

4 a Prove algebraically that the sum of four consecutive numbers is not divisible by 4.

b What generalisation can you make about the sum of four consecutive numbers?

5 Prove algebraically that if n is an even number, then n^2 must also be even.

Tip

All even numbers can be written in the form of $2m$, where m is any whole number.

6 If two numbers a and b are even integers, prove that ab is divisible by 4.

7 Prove algebraically that the sum of any three consecutive even numbers is a multiple of 6.

4 Functions and sequences

Section 1: Sequences and patterns

HOMEWORK 4A

1 Find the next three terms in each sequence. Describe the rule you used to find them.
   a 11, 13, 15 …
   b 88, 99, 110 …
   c 64, 32, 16 …
   d 8, 16, 24 …
   e ⁻2, ⁻4, ⁻6, ⁻8 …
   f $\frac{1}{4}, \frac{1}{2}, 1$ …
   g 1, 2, 4, 7 …
   h 1, 6, 11, 16 …

2 List the first four terms of the sequences that follow these rules.
   a Start with 7 and add 2 each time.
   b Start with 37 and subtract 5 each time.
   c Start with 1 and multiply by $\frac{1}{2}$ each time.
   d Start with 5 then multiply by 2 and add 1 each time.
   e Start with 100, divide by 2 and subtract 3 each time.

Tip

Look at how the numbers change each time. Is the change the same?

3 Josh skims a stone across a pond.
   Each 'bounce' is $\frac{2}{3}$ the length of the previous one.
   a If the first bounce is 216 cm, how long will the fourth bounce be?
   b How many times will the stone bounce before the bounce is less than 1 cm?

4 Find the term-to-term rule for each of these sequences and use it to give the next three numbers in each sequence.
   a 4.3, 5.3, 6.3   b 0.7, 1.4, 2.1   c $\frac{3}{4}, 1\frac{1}{4}, 1\frac{3}{4}$
   d 7, 4, 1   e 28, 14, 7   f ⁻8, ⁻13, ⁻18

Section 2: Finding the *n*th term

HOMEWORK 4B

1 Find the expressions for the *n*th term in the following sequences.
   a 5, 9, 13, 17…
   b 3, 5, 7, 9…
   c 3, 7, 11, 15…
   d ⁻1, 4, 9, 14…
   e 7, 12, 17, 22…
   f ⁻3, 0, 3, 6…
   g ⁻1, 6, 13, 20…

Tip

Start by finding the difference between the terms.

2 Bonita is conducting an experiment in science and gets the following pattern of results:
   53, 61, 69, 77, …
   Write an expression for the nth term for Bonita's results.

3 Consider the sequence:
   2, 10, 18, 26, 43, 42, 50 . . .
   a Find the *n*th term of the sequence.
   b Find the 200th term.
   c Which term of this sequence has the value 234? Show full working.
   d Show that 139 is not a term in the sequence.

4 For each sequence below find the general term and the 50th term.
   a 7, 9, 11, 13 . . .
   b ⁻5, ⁻13, ⁻21, ⁻29 . . .
   c 2, 8, 14, 20, 26 . . .
   d 4, 9, 16, 25 . . .
   e 2.3, 3.5, 4.7, 5.9 . . .

5 Find the expressions for the nth term of the following sequences.
   a 29, 25, 21, …   b ⁻15, ⁻22, ⁻29, …
   c 6.25, 5.45, 4.65, …

Section 3: Functions
HOMEWORK 4C

1. The numbers 1 to 10 are put in order into the function $T(n) = n - 2$.
   What sequence does this create?

2. What sequences would be created by putting the numbers 1 to 10 in order into the following functions:
   a $T(n) = n + 4$
   b $T(n) = 4n$
   c $T(n) = n - 7$
   d $T(n) = \dfrac{n}{3}$
   e $T(n) = 2n + 1$

3. What sequences would be created by putting the numbers $^-5$ to 5 in order into the following functions?
   a $T(n) = 3n$
   b $T(n) = 2n + 3$
   c $T(n) = 3n - 4$
   d $T(n) = 2n + \dfrac{1}{4}$
   e $T(n) = \dfrac{n}{2} - 1$

4. Write an equation for the composite function formed by each of these pairs of functions:
   i function 1 = $3n$, function 2 = $n - 6$
   ii function 1 = $5n$, function 2 = $n + 7$
   iii function 1 = $7n$, function 2 = $n - 4$
   iv function 1 = n^2, function 2 = $n + 3$

   > **Tip**
   >
   > Remember, a composite function is the combination of two or more functions.

5. Write down the function that is the inverse of each function given below.
   a $T(n) = n - 7$ b $T(n) = 4n$
   c $T(n) = n + 5$ d $T(n) = \dfrac{n}{3}$

   > **Tip**
   >
   > Remember, a function that reverses the result of a different function is called the 'inverse function'.

Section 4: Special sequences
HOMEWORK 4D

1. If a cow produces its first female calf at age two years and after that produces another single female calf every year, how many female calves are there after 12 years, assuming none die and that each cow produces calves in the same way?

2. Write down the sequence of the first 12 triangle numbers, with each term doubled.

3. a Following the rules for the Fibonacci series, but starting with the numbers $^-5$ and 3, write down the first ten terms of the sequence.
   b Following the same rule, start the sequence with the numbers 3 and $^-5$. Write down the first ten terms of the sequence. Does this generate the same sequence?

4. The 7th and 8th terms of a sequence formed using the Fibonacci rule are $^-14$ and $^-23$. What are the first two terms?

5. Write down the first ten terms of the sequence formed by using the rule $2n^2 - n$.

6. Write down the first ten terms of the sequence formed by $\dfrac{(n^2 + n)}{2}$.
   What is the name given to this sequence?

HOMEWORK 4E

1. Find an expression for the nth term for each of these sequences.
   a 5, 10, 17, 26, 37
   b 4, 11, 22, 37, 56
   c 5, 20, 43, 74, 113
   d 9, 20, 35, 54, 77
   e $^-4$, 5, 22, 47, 80
   f $^-4$, $^-10$, $^-20$, $^-34$, $^-52$

   > **Tip**
   >
   > You will need the second difference to find the nth term of a quadratic sequence.

2. A sequence is defined as $U_n = n^2 - 2n + 4$. Write the first five terms of this sequence.

3. Find the next five terms in this sequence:
   $1, \sqrt{3}, 3, 3\sqrt{3},$

Chapter 4 review

1 A new bacteria is growing in a laboratory. After 1 hour it consists of 10 cells, 16 after 2 hours, 22 after 3 hours and 28 after 4.

  a If it continues to grow the same rate, how many cells will there be after 24 hours?

  b Find an expression that will work out the number of cells after any number of hours.

2 The number of rats on an island is recorded each month.

After 1 month there are 8 rats.

After 2 months there are 20 rats

After 3 months there are 32 rats

  a If the population keeps growing at the same rate, and no rats die, how many rats will there be at the end of the year?

  b Explain why a sequence of this type is unlikely to work in reality.

3 A basketball is dropped from a height of 16 m. For each bounce the ball returns to $\frac{3}{4}$ of the height of the previous bounce.

How high will the 5th bounce be?

Give your answer to the nearest centimetre.

4 What sequences would be created by putting the numbers 1 to 10 in order into the following functions:

  a $T(n) = 3 + n$ **b** $T(n) = {}^-2n$

  c $T(n) = 6 - n$ **d** $T(n) = \dfrac{n}{4}$

  e $T(n) = 3n - 1$

5 For the sequence $2n - 3$, write down:

  a the first five terms   **b** the 10th term

  c the 25th term.

6 Find the expressions for the nth term in the following sequences:

  a 4, 6, 8, 10, ...   **b** 5, 9, 13, 17, ...

  c $^-$2, 1, 4, 7, ...   **d** 2, 7, 12, 17, ...

7 Write down the first ten terms of the sequence formed by $\dfrac{(n^2 + 1)}{2}$.

8 The number of hits of a video on a website is increasing day by day.

On day one there were 16 hits, on day two 30 hits, day three 58 hits and day four 100 hits. Assuming the number of hits continues to grow at the same rate, how many hits will there be on day ten?

5 Properties of shapes and solids

Section 1: Types of shapes

HOMEWORK 5A

1 What is the correct mathematical name for each of the following shapes?

  a A plane shape with four sides.

  b A polygon with six equal sides.

  c A polygon with five vertices and five equal angles.

  d A plane shape with ten equal sides and ten equal internal angles.

Tip

Learn the names of shapes and be clear which are regular and which are irregular.

2 What are the names of the following quadrilaterals?

  a   **b**   **c**

Tip

It is a good idea to learn the properties of quadrilaterals.

3 Name each shape, given the following properties.
   a A four-sided shape with two pairs of equal and opposite sides but no right angles.
   b A four-sided shape with only one pair of parallel sides.
   c A triangle with two equal angles.
   d A triangle with all sides and angles equal.
   e A four-sided shape with two pairs of equal and adjacent sides.

HOMEWORK 5B

1 Look at this diagram. Say whether the following statements are true or false.

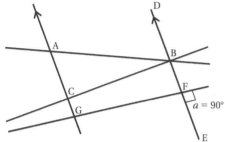

   a AG is parallel to DE.
   b ABC is an isosceles triangle.
   c DE is perpendicular to BC.
   d AG is perpendicular to GF.
   e AB is perpendicular to AG.
   f AB and GF are parallel.

2 Draw and correctly label a sketch of each of the following shapes.
   a Triangle ABC with a right angle at A and AB = AC.
   b A quadrilateral PQRS with two pairs of opposite equal angles, none of which are right angles, and two pairs of opposite equal sides.
   c Quadrilateral ABCD where AB is parallel to CD and the angle ABC is a right angle.

Section 2: Symmetry
HOMEWORK 5C

1 How many lines of symmetry do the following shapes have?
   a A square   **b** A kite
   c A regular hexagon
   d An equilateral triangle

> **Tip**
>
> Line symmetry cuts a shape in half so that one side is a mirror image of the other.

2 Give an example of a shape that has rotational symmetry of order:
   a 2   **b** 3   **c** 4

3 Which of the following letters have rotational symmetry?

N I C K

> **Tip**
>
> Rotational symmetry is when the shape looks exactly the same after a rotation.

4 What is the order of rotational symmetry for each of these images?

a    **b**

c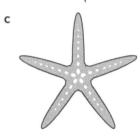

Section 3: Triangles
HOMEWORK 5D

> **Tip**
>
> Learn the properties of the different types of triangle.

1 What type of triangle can be seen below? Explain how you decided without measuring.

11

2 **a** What type of triangle is this?

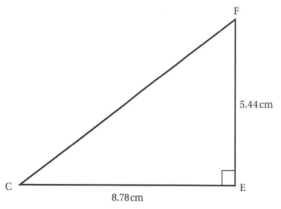

b Explain why this triangle cannot be isosceles.

3 State whether the following triangles are possible. How did you decide?
  a Side lengths 6 cm, 8 cm, 10 cm
  b Side lengths 12 cm, 4 cm, 5 cm
  c Side lengths 7 cm, 11 cm, 5 cm
  d Side lengths 35 cm, 45 cm, 80 cm

4 Two angles in a triangle are 27° and 126°.
  a What is the size of the third angle?
  b What type of triangle is this?

Tip

Use the properties of triangles and angles to answer this question.

5 Look at the diagram below.

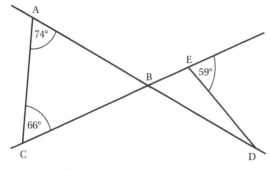

Work out the following:
  a Angle ABC
  b Angle BED
  c Angle BDE

6 An isosceles triangle PQR with PQ = QR has a perimeter of 80 cm. Find the length of PQ if:
  a PR = 24 cm
  b PR = 53 cm.

7 In the diagram below, the line CD is parallel to AB.
Find the following angles, giving reasons for your answers.
  a Angle ABD.
  b Angle CED.
  c Angle BDC.

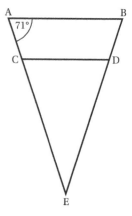

8 The triangle ABC is isosceles and has two angles of $2x - 5$ and one of $2x - 10$. Calculate the size of each angle.

9 The triangle PQR has an angle PQR = $6x + 7$ and QRP = $4x + 8$.
The angle RPQ = $2 \times$ angle QRP.
Find the value of x and hence the size of each angle in the triangle.

Section 4: Quadrilaterals
HOMEWORK 5E

1 Identify the quadrilateral from the description. There may be more than one correct answer.
  a All sides are equal.
  b Diagonals cross at right angles.
  c One pair of sides is parallel.
  d Two pairs of sides are parallel and equal in length.

Tip

Make sure you learn the names and properties of all the quadrilaterals.

2 Molly says that all four-sided shapes have at least one pair of equal or parallel sides.
Is she right?

3 A kite ABCD has angle ABC 43° and the opposite angle ADC 75°.
What size are the other two angles?

4 One pair of triangles has the angles 36°, 54° and 90°, while another pair has the angles 24°, 66° and 90°. The length of the shortest side in each of the four triangles is the same.

Imagine all four triangles placed together so that the right angles meet at the same point.
a What shape has been formed?
b What are the sizes of the angles at the vertices of this new shape?

5 **a** Write down the names of all the different quadrilaterals.
b Which quadrilaterals have at least two equal sides?
c Which quadrilaterals have at least one pair of parallel sides?
d Which quadrilaterals have no rotational symmetry?

6 The diagram shows a rhombus.

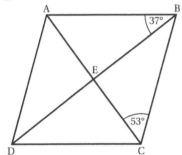

Calculate.
a Angle BAE
b Angle EBC
c Angle EDA
d Angle ADC

7 The quadrilateral ABCD is a parallelogram such that AB is parallel to CD and AD is parallel to BC.
If the angle DAB = $4x + 7$ and the angle ADC = $2x + 8$, find the value of x and hence the size of the angles DAB and ADC.

8 The quadrilateral PQRS has diagonals SQ and PR that meet at point T.
The angle SPT $- 3x$, RSP $- 4x$, QRS $- x$, PQR = $4x$ and TPQ = $3x$.
Find the value of x and hence the size of the angles SPQ, PQR, QRS and RSP.
What shape is PQRS?

Section 5: Properties of 3D objects
HOMEWORK 5F

1 Imagine a solid made by fixing six square-based pyramids to a cube when the squares of the pyramids have the same side length as that in the cube. The edges of the cube and the edges of the pyramids are aligned.
How many vertices, edges and faces would this solid have?

2 How many vertices, edges and faces do the following have?
a Tetrahedron   **b** Hexagonal prism
c A prism with an L-shaped cross-section

3 A rectangular-based pyramid is sliced parallel to the base.
Describe the two shapes created.

Chapter 5 review

1 True or false?
a A triangle with two equal angles is called isosceles.
b A cuboid has 8 vertices, 6 faces and 10 edges.
c A pair of lines that meet at precisely 90° are described as being perpendicular.
d Every square is a rhombus.
e Every rectangle is a parallelogram.
f Every square is a rectangle.

2 Describe all the symmetrical features of a rectangle.

3 Find the size of the missing angles in this trapezium.

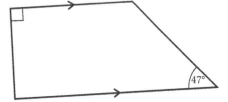

4 In the diagram below AC is parallel to FD and AE is parallel to BD. FB and EC are also parallel. Find the missing angles. Give reasons for your answer.

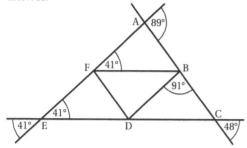

a Angle FAB
b Angle ABF
c Angle BDC
d Angle FBD

5 Draw a diagram and prove that the exterior angle of a triangle is equal to the sum of the two opposite interior angles.

6 Prove that a triangle formed by drawing a straight line parallel to one arm of an angle from any point on the bisector of the angle is isosceles.

7 A point P is drawn on the side BC of an isosceles triangle ABC where AB = AC. A line perpendicular to BC is drawn through P that cuts the side AB at Q. The side AC meets the perpendicular line at R.
Prove that the triangle QAR is isosceles.

6 Construction and loci

Section 1: Geometrical instruments
HOMEWORK 6A

1 Use a ruler and protractor to draw and label the following angles.
a ABC = 35°
b DEF = 129°
c PQR = 100°

> **Tip**
>
> A sharp pencil and a good quality compass that has been tightened are essential to construct accurately.

2 Explain how to use a protractor marked from 0° to 180° to measure a reflex angle.

3 Draw line AB, which is 6.2 cm long.
At A, measure and draw angle BAC = 45°.
At B, measure and draw angle ABD = 98°.

4 Use a pair of compasses to construct:
a a circle of radius 3 cm
b a circle of diameter 10 cm
c two circles, one of diameter 3 cm, and another of diameter 4 cm, whose circumferences touch once.

5 Draw a circle of radius 45 mm and centre O.
Use a ruler to draw any two radii of the circle. Label them OA and OB.
Join point A to point B to form triangle AOB.
Measure the angles AOB, OBA and BAO.
What sort of triangle have you constructed?

6 Use a ruler and a pair of compasses to construct $\frac{1}{6}$ of a circle of diameter 72 mm.

Section 2: Bisectors and perpendiculars
HOMEWORK 6B

1 Measure and draw the following line segments. Find the midpoint of each by construction.
a AB = 11 cm
b CD = 36 mm
c EF = 7.5 cm

> **Tip**
>
> Make sure you know how to find a midpoint by construction.

2 Using ruler and compasses only, draw an angle of 60°. Bisect the angle without measuring. Check your accuracy by measuring.

> **Tip**
>
> Think of a shape that has angles of 60°.

3 Draw the following angles and then, using only a ruler and pair of compasses, bisect each angle.
  a An angle of 68°.
  b An angle of 156°.
  c An angle of 120°.

4 Draw a triangle ABC where AB = 6 cm, BC = 7 cm and AC = 8 cm.
  a Construct the bisector of each angle.
  b Use the point where the bisectors meet as the centre and draw a circle whose radius is the shortest distance to the side of the triangle.
  c What do you notice about this circle?

5 Draw AB = 90 mm. Insert any point C above AB.
  a Construct CX perpendicular to AB.
  b Draw CD parallel to AB.

6 a Draw a circle of radius 60 mm.
  b Construct a chord that is 24 mm long.

> **Tip**
>
> Use a pair of compasses.

  c Join both ends of the chord to the centre of the circle. Bisect the angle formed at the centre. What can you say about the intersection between the chord and the angle bisector?

Section 3: Loci
HOMEWORK 6C

1 Sketch the point, path or area that each locus will produce.
  a Points that are 50 m from a flagpole at point X.

  b The area of grass a goat can eat if tethered to the corner of a rectangular field where the length of rope is the same length as the short side of the rectangle.
  c Points that are equidistant from each track of a single railway line.

2 Accurately construct the locus of points 6 cm from a point A.

> **Tip**
>
> A locus is a set of points that satisfy the same rule.

3 Draw angle ABC = 70°.
Accurately construct the locus of points equidistant from AB and BC.

> **Tip**
>
> Remember to leave your construction arcs.

4 Draw PQ 50 mm long.
Construct the locus of points 2 cm from PQ.

5 Draw a rectangle ABCD with AB = 7 cm and BC = 5 cm.
  a Shade the locus of points that are closer to AB than CD and within the rectangle.
  b Shade the locus of points that are less than 2 cm from A and within the rectangle.
  c Construct the locus of points that are equidistant between AD and BC and within the rectangle.

6 MNOP is a square with sides of 6 cm. Show by construction the locus of all points that are less than 1 cm from the sides of the square and which are outside the square.

7 Draw a diagram to show the locus of a windscreen wiper as it wipes a car window. Assume the windscreen is rectangular, that there is one wiper blade connected to the midpoint of the base of the rectangle.

Section 4: More complex problems
HOMEWORK 6D

1 Draw line PQ = 6.4 cm.
Construct RS, the perpendicular bisector of PQ.
Draw RS so that it is 6.4 cm long and the mid-point of RS is at the same point as the mid-point of PQ.
Construct the locus of points that are 3.2 cm from each of P, Q, R and S.

Tip

These problems involve careful and accurate construction. Make sure you use a ruler and compasses.

2 Construct a parallelogram with sides of 5.5 cm and 3.2 cm and a longest diagonal of length 7 cm. How long is the other diagonal?

3 Accurately construct a square of side 62 mm.

4 Construct quadrilateral PQRS such that PSR is a right angle, SRQ is a right angle, QR = 2PS and SR = 5 cm.
What kind of quadrilateral is this?

5 Two towns C and D lie 6 km apart. Two TV transmitters P and Q lie 5 km apart, equidistant from the line CD. The line PQ is perpendicular to CD and 2 km from C. Each transmitter can transmit 3.5 km in any direction.
a Draw a diagram to show the range of the transmitters.
b If the range of the transmitters was increased to 4.5 km, would the residents of D be able to receive TV signals?

6 A rectangular courtyard 70 m by 80 m has a CCTV camera in each corner. The effective range of each camera is 45 m.
a Draw a scale diagram to show the extent of the courtyard that is covered by CCTV.
b What would be the minimum range of the cameras required to ensure that all of the courtyard is covered?

7 Town A is 8 km from B and 9 km from C. B and C are 7.5 km apart.
The council wishes to connect the towns using the shortest road distance possible.
Draw a scale diagram to show how this might be achieved. Show the lengths of each road on your diagram.

Chapter 6 review

1 **a** Draw an angle of precisely 68°.
b Bisect this angle by construction.

2 Draw a line PQ, which is 45 mm long.
Use this line as the diameter of a circle.

3 Draw a line AB of length 8.4 cm and find its midpoint by construction.
Show the locus of points that are equidistant from A and B on your diagram.

4 Draw accurately a circle of circumference of 44 cm.
Draw a radius of this circle.
Draw the loci of all points that are within 15 mm of the radius.

7 Further algebraic expressions

Section 1: Multiplying two binomials
HOMEWORK 7A

 1 Expand and collect like terms.
- **a** $(x + 1)(x + 4)$ **b** $(x + 3)(x + 5)$
- **c** $(a + 5)(a + 4)$ **d** $(6 + x)(3 + x)$
- **e** $(7 + x)(x + 2)$ **f** $(a + 6)(7 + a)$

> **Tip**
>
> Each term in a bracket must be multiplied by each term in the other bracket.

2 Find these products and simplify.
- **a** $(x - 4)(x - 2)$ **b** $(a - 6)(a - 3)$
- **c** $(m + 3)(m - 6)$ **d** $(p - 7)(p + 5)$
- **e** $(x - 8)(x + 5)$ **f** $(x + 12)(x - 2)$

3 Expand and simplify.
- **a** $(2x + 5)(2x + 4)$ **b** $(2x + 3)(5x + 4)$
- **c** $(2x - 3)(4x + 7)$ **d** $(4x - 6)(6x + 3)$
- **e** $(4x - 7)(2x - 2)$ **f** $(3x - 7)(x - 4)$

HOMEWORK 7B

1 Expand each of these perfect squares.
- **a** $(x + 3)^2$ **b** $(x + 5)^2$ **c** $(x - 4)^2$
- **d** $(x - 11)^2$ **e** $(2x + 3)^2$ **f** $(2 - 4x)^2$

2 Expand each of the following binomials.
- **a** $(a + 1)(a - 1)$ **b** $(x + 3)(x - 3)$
- **c** $(2x + 1)(2x - 1)$ **d** $(2x - y)(2x + y)$

HOMEWORK 7C

1 Expand the brackets and simplify.
- **a** $(x - 5)(2x + 3)(x + 7)$
- **b** $(2x - 1)(x + 3)(x + 4)$

2 Expand the brackets and simplify.
- **a** $\left(\frac{3x}{2} + 5\right)\left(\frac{x}{4} - 1\right)$ **b** $(3x - 2)^2 + (x - 3)^2$
- **c** $x^2 + (x + 1)^2 + (x + 2)^2 + (x + 3)^2$
- **d** $(x - 2)(2x^2 + x + 3)$
- **e** $(x - 1)(x - 2)(x - 3)(x - 4)$

3 The volume of a cuboid can be found using the formula LWH, where L is the length, W is the width and H is the height.

Given a cuboid of length $\left(3x + \dfrac{1}{2}\right)$ cm, width $(2x - 1)$ cm and height $(x - 3)$ cm:
- **a** Write an expression for the volume of the cuboid in factor form.
- **b** Expand the expression.

Section 2: Factorising quadratic expressions
HOMEWORK 7D

1 Fill in the blank boxes.
- **a** $(x + 6)(x + 8) = x^2 + \boxed{}x + \boxed{}$
- **b** $(x + \boxed{})(x + 6) = x^2 + 10x + \boxed{}$
- **c** $(x + 7)(x - \boxed{}) = x^2 - 2x - \boxed{}$
- **d** $(3x - 2)(x - \boxed{}) = 3x^2 - \boxed{} + 6$
- **e** $(\boxed{}x + \boxed{})(2x + 7) = 4x^2 + 18x + \boxed{}$

2 Find two numbers that meet each set of conditions.
- **a** Have a sum of 7 and a product of 12.
- **b** Add to give 8 and multiply to give 12.
- **c** Have a product of $^-14$ and a sum of 5.
- **d** Multiply to give 36 and add to give $^-13$.

3 Factorise these quadratic expressions.
- **a** $x^2 + 7x + 12$
- **b** $x^2 + 5x + 4$
- **c** $x^2 + 11x + 30$

> **Tip**
>
> Factorising is the opposite of expanding.

 4 Factorise these quadratic expressions.
- **a** $x^2 - 6x + 8$
- **b** $x^2 - 6x + 5$
- **c** $x^2 - 8x + 12$

5 Factorise these quadratic expressions.
   a $x^2 + x - 6$ **b** $x^2 + 4x - 5$ **c** $x^2 - 3x - 10$

6 Factorise fully.
   a $4x^2 + 12x + 8$ **b** $6x^2 - 18x - 24$
   c $2x^2 + 6x - 20$

HOMEWORK 7E

1 Factorise each expression.
   a $x^2 - 9$ **b** $x^2 - 36$ **c** $x^2 - 121$

2 Using $(a - b)(a + b) = a^2 - b^2$, evaluate the following.
   a $80^2 - 76^2$ **b** $48^2 - 37^2$ **c** $754^2 - 749^2$

3 Factorise fully.
   a $6x^2 - 54$ **b** $5x^2 - 80$
   c $3x^2 - 75$ **d** $7x^2 - 7$

HOMEWORK 7F

1 Factorise completely.
   a $2x^2 + 4x - 6$ **b** $4x^2 - 12x + 8$
   c $3x^2 + 15x + 12$ **d** $4x^2 + 8x - 60$

2 If the area of a parallelogram is $6x^2 - x - 2$ and the length is $3x - 2$, find the perpendicular height.

3 The area of an isosceles triangle is $20x^2 + 11x - 3$ and the perpendicular height is $5x - 1$. Find the length of the base.

Section 3: Completing the square
HOMEWORK 7G

1 Expand these perfect squares.
   a $(x + 6)^2$ **b** $(y - 4)^2$
   c $(3x + 5)^2$ **d** $\left(3x - \dfrac{3}{4}\right)^2$

2 Factorise using the difference of two squares rule.
   a $x^2 - 4$ **b** $16x^2 - 1$
   c $16x^2 - \dfrac{1}{4}$ **d** $3x^2 - 18$

3 Find the missing terms in these trinomials to make them equivalent to the perfect square.
   a $x^2 + 10x + \boxed{} = (x + 5)^2$
   b $x^2 - 14x + \boxed{} = (x - 7)^2$
   c $x^2 + 7x + \boxed{} = \left(x + \dfrac{7}{2}\right)^2$

d $x^2 + 11x + \boxed{} = \left(x + \dfrac{11}{2}\right)^2$

e $x^2 + 5x + \boxed{} = \left(x + \dfrac{5}{2}\right)^2$

4 Complete the square.
   a $x^2 + 2x - 7$ **b** $x^2 + 4x + 9$ **c** $x^2 + 6x + 4$
   d $x^2 + 10x - 3$ **e** $x^2 - 6x + 7$ **f** $x^2 - 10x - 5$

Section 4: Algebraic fractions
HOMEWORK 7H

1 Simplify.
   a $\dfrac{(4x + 2)}{(2x + 1)}$ **b** $\dfrac{(x^2 - 4)}{(x - 2)}$
   c $\dfrac{(9x^2 - 16)}{(3x - 4)}$ **d** $\dfrac{(x + 3)(x - 2)}{(x + 1)(x + 3)}$

2 Express as single fractions.
   a $\dfrac{3x}{2} + \dfrac{2x}{3}$ **b** $\dfrac{5}{3x} + \dfrac{2}{5x}$
   c $\dfrac{1}{(x + 3)} + \dfrac{1}{(x + 2)}$ **d** $\dfrac{9}{(x - 3)} - \dfrac{3}{(x + 2)}$

3 Express as single fractions.
   a $\dfrac{x - 1}{x + 2} - \dfrac{x - 3}{x + 4}$ **b** $\dfrac{x + 5}{x - 1} + \dfrac{2}{x + 1}$
   c $\dfrac{1}{3x - 2} + \dfrac{1}{3x + 2}$ **d** $\dfrac{9}{(x - 3)} - \dfrac{3}{(x + 2)}$

Section 5: Apply your skills
HOMEWORK 7I

1 Show algebraically that $(x + 1)$ is not a factor of $(2x^2 + 13x + 15)$.

2 Given that $A = 2x + 3$ and $B = 3x - 1$, write each of the following expressions in terms of x in their simplest form.
   a AB **b** $A^2 + B^2$
   c $(A - B)(A + B)$

3 A square is transformed into a rectangle by subtracting 2 cm from one side and doubling the other side and subtracting 5 cm from it.
   a Write an expression for the area of the rectangle
   b If the original square side length was 20 cm, what is the difference between the area of the rectangle and the area of the original square?

4 Use the method of completing the square to show that:

a $x^2 + 6x + 21 \geqslant 12$ b $x^2 + 10x + 35 \geqslant 10$

5 Use $a^2 - b^2 = (a + b)(a - b)$ to evaluate $2001^2 - 1999^2$.

6 The area of a quadrilateral is expressed as $x^2 + 12x + 36$
Can this shape be a square?

Chapter 7 review

1 Expand and simplify.

a $(x - 1)^2 - (x - 2)^2 - (x - 3)^2$
b $(x + 2)(x - 3)(x + 4)$

2 Factorise.

a $9x^2 - 18x - 72$ b $4x^2 + 12x + 9$

3 Use the difference of two squares identity to calculate $921^2 - 919^2$.

4 The diagram shows a trapezium. The lengths of three of the sides of the trapezium are shown. All dimensions are in centimetres.

(Diagram: trapezium with sides labelled $3x + 5$, $x + 1$, $2x - 1$)

a Write an expression for the area of the trapezium.
b If $x = 5$ cm, find the area of the trapezium.

5 Write as single fractions in simplest form.

a $\dfrac{x + 1}{2} + \dfrac{x - 3}{3}$ b $\dfrac{1}{x + 3} + \dfrac{3}{x + 2}$

c $\dfrac{3x}{2x + 1} - \dfrac{2x}{x - 1}$ d $\dfrac{4}{x + y} - \dfrac{2}{x - y}$

e $\dfrac{1}{x + 3} + \dfrac{3}{x + 3} + \dfrac{5}{x - 2}$

8 Equations

Section 1: Linear equations
HOMEWORK 8A

1 Solve these equations.

a $x + 6 = 9$ b $x - 5 = 12$

c $^-3x = 18$ d $x + 4 = 22\frac{1}{2}$

e $4x = {}^-16$ f $5x - 21 = 9$

> **Tip**
>
> When solving a linear equation you are trying to find the value of the letter that will make the equation true.

2 Solve these equations.

a $2a - 5 = 1$ b $5b + 4 = 24$

c $7d - 7 = 42$ d $5e - 2 = 28$

e $12h + 11 = 35$ f $3x + 13 = {}^-5$

g $^-8x - 5 = {}^-53$ h $7x - 52 = {}^-87$

3 Solve these equations.

a $2(x + 5) = 16$ b $4(x - 3) = 8$
c $7(x - 2) = 21$ d $4(x - 3) = {}^-20$
e $2(x - 5) = 42$ f $^-3(x - 4) = 27$

HOMEWORK 8B

1 Solve the following equations. Check by substitution.

a $3x - 7 = 2x + 5$ b $4x + 3 = 5x - 2$
c $5x - 6 = 6x - 17$ d $2x + 10 = 5x + 25$
e $6x - 9 = 8x + 5$ f $4x - 11 = 6x - 1$

2 Solve these equations by expanding the brackets first.

a $2(x + 4) = 11(x - 5)$ b $3(x + 3) = 6(x - 6)$
c $5(x - 2) = \frac{1}{2}(x + 7)$ d $2(x - 2) = 4(x - 4)$
e $3(x - 2) = 4(x - 1)$ f $\frac{1}{2}(x + 7) = 6(x - 4)$

3 Which of these pairs of equations are identities? Change = to ≡ if you think it is an identity.
a $3(x + 4) - 5x = 2x + 4$
b $6(x + 3) - 4x = 2(x + 9)$
c $2(x - 3) + 6x - 4 = 8x - 7$

4 Solve for x.

Tip

Use the rules for adding and subtracting fractions as you would with numbers.

a $\dfrac{x + 10}{7} + \dfrac{2x}{5} = 8$ **b** $\dfrac{4}{x + 1} = \dfrac{5}{x - 2}$

c $\dfrac{x}{3} + \dfrac{x}{2} = 10$ **d** $\dfrac{5x - 2}{3} = \dfrac{12x + 3}{7}$

e $\dfrac{-2(5 + x)}{3} = {}^-11$ **f** $\dfrac{2x - 13}{4} - \dfrac{2x}{3} = 5 + 2x$

HOMEWORK 8C

1 For each of the following, write an equation and solve it to find the unknown number.
a Four times a certain number is 212. What is the number?
b 6 less than a number is ⁻5. What is the number?
c 11 greater than a number is ⁻6. What is the number?
d Three less than six times a number is 39. What is the number?
e Two consecutive numbers have a sum of 43. What are the numbers?

2 A rectangle has the side lengths $x + 4$ and $2(x - 1)$.
a Write an expression for the perimeter of the rectangle.
b Find the value of x if the perimeter is 40 cm.

3 I have three piles of stones. The second pile has twice as many as the first pile, and the third pile has four more than the second pile. Altogether I have 64 stones. How many stones in each pile?

4 Wilf buys 20 stamps and gets £1.60 change from £10. How much do the stamps cost each?

5 Multiplying a certain number by six and adding 11 to the result gives the same answer as multiplying the number by eight and subtracting five from the result. What is the number?

Section 2: Quadratic equations
HOMEWORK 8D

Tip

In a quadratic equation the largest power of a variable is squared.

1 Solve for x.
a $x^2 - 6x = 0$ **b** $x^2 + x = 0$
c $5x^2 + x = 0$ **d** $3x^2 + x = 0$

2 Solve for x.
a $x^2 - 25 = 0$ **b** $81 - x^2 = 0$
c $9x^2 - 1 = 0$ **d** $4x^2 - 36 = 0$

3 Find the roots of each equation.
a $x^2 + x - 6 = 0$ **b** $x^2 - 7x + 12 = 0$
c $4x^2 + 4x - 48 = 0$ **d** $6x^2 - 18x + 12 = 0$

4 Solve these equations.
a $x^2 + 6x = {}^-8$ **b** $x^2 + 3x = 10$
c $x^2 - 2x - 2 = 13$ **d** $x^2 - 6x = {}^-8$
e $6x^2 + 7x = 2$ **f** $5x^2 - 9x + 1 = 3$

5 Solve these equations.
a $x^2 - 2x = 15$ **b** $x^2 - 6x - 2 = {}^-6$
c $3x^2 + 3x = 36$ **d** $5x^2 - 15x - 1 = 19$

HOMEWORK 8E

1 Solve for x. Give your answers in surd form.
a $(x + 2)^2 = (x + 1)^2 + 16$

b $\dfrac{5x + 3}{4} = \dfrac{9}{x}$ **c** $\dfrac{3}{2x - 1} = \dfrac{x}{5x - 2}$

2 Solve these equations by completing the square.
a $x^2 + 12x + 20 = 0$ **b** $x^2 - 6x - 8 = 0$
c $x^2 - 4x + 1 = 0$ **d** $x^2 + 4x - 1 = 0$

3 Use the quadratic formula to solve each quadratic equation, giving your answers to two decimal places where appropriate.
a $x^2 - 4x - 12 = 0$ **b** $x^2 + 8x + 9 = 0$
c $2x^2 - 16x + 30 = 0$ **d** $4x^2 - 4x - 120 = 0$

4 Use the quadratic formula to solve each quadratic equation. Give your answers in simplest surd form.

a $x^2 + 10x + 7 = 0$ b $3x^2 + 4x - 9 = 0$

c $4x^2 + 7x - 2 = 0$ d $5x^2 - 8x + 3 = 0$

HOMEWORK 8F

1 Form an equation and solve it to find the unknown numbers.

a The product of a certain positive whole number and three more than that number is 270.
What could the number be?

b The product of a certain positive whole number and five less than that number is 126.
What could the number be?

c The difference between the square of a number and twice the original number is 8.
What are possible values of the number?

d The product of two consecutive positive even numbers is 168.
What are the numbers?

2 The base of a triangle is 6 cm longer than twice its height.
If the area of the triangle is 54 cm², calculate its height.

Section 3: Simultaneous equations
HOMEWORK 8G

Tip

Simultaneous equations have the same solution. The solution **must** satisfy both equations.

1 Solve the following pairs of simultaneous equations by substitution.

a $x + y = 3$ b $x - y = 4$
  $2x + y = 4$ $3x + y = 8$

c $x + y = 2$ d $3x + 2y = 23$
  $3x - y = 14$ $x - 2y = 1$

e $2x + 2y = 6$ f $3x + y = 12$
  $2x + y = 0$ $2x - y = 18$

2 Solve these simultaneous equations.

a $2x + y = 10$ b $3x + 2y = {}^-4$
  $3x + 2y - 16$ $x - 2y - 12$

c $5x + 2y = {}^-11$ d ${}^-3x + 2y = {}^-13$
  $6x - 2y = 0$ $3x + 4y = 37$

e $3x + 2y = {}^-33$ f $5x - 3y = 2$
  $4x - 4y = {}^-24$ $4x - 6y = {}^-20$

HOMEWORK 8H

1 Solve the following pairs of simultaneous equations by elimination.

a $2x + 3y = {}^-13$ b $3x + y = {}^-37$
  $4x - 2y = {}^-50$ $5x - 4y = {}^-22$

c $3x + 5y = 37$ d $6x - 2y = 28$
  $4x + 2y = 26$ $4x - 3y = 32$

e $3x + 4y = 39$ f $5x - 4y = {}^-32$
  $9x + 3y = 36$ $4x + 5y = 40$

2 Solve each pair of simultaneous equations. Choose the most suitable method for doing this.

a $x + y = 0$ b $2x - y = 3$
  $3x + 2y = 2$ $4x + y = 21$

c $3x + 2y = {}^-3$ d $x - 3y = 10$
  $2x - 3y = {}^-12$ $2x + y = {}^-1$

e $6x + 2y = {}^-28$ f $5x + 8y = {}^-2$
  $2x - 3y = {}^-2$ $6x - 3y = 48$

3 Solve each pair of simultaneous equations by substitution.

a $y = x^2$ b $y = x^2$
  $y = x + 2$ $y = 4x - 3$

c $y = x^2$ d $x^2 + y^2 = 25$
  $y = 2x + 3$ $y = x + 1$

HOMEWORK 8I

1 Ben and Molly buy their friends drinks at the café.
Ben bought three shakes and two coffees for £10.20.
Molly bought four shakes and one coffee for £10.35.
What is the cost of:

a one shake?   b one coffee?

2 Shazan is counting the money in her till. She has 37 notes, some of which are £5 and some £10, making a total of £280.
How many of each note does she have?

3 Two numbers have a sum of 62 and a difference of 24.
What are the numbers?

4 The sum of two numbers, x and y, is 80.
When $2y$ is subtracted from $5x$, the result is 99.
Find the value of x and y.

5 A plumber charges a call-out fee plus an amount per hour.
A job taking 5 hours costs £155 and a job taking $3\frac{1}{2}$ hours costs £117.50.
How much would a job taking 8 hours cost?

HOMEWORK 8J

1 Find the coordinates of the point(s) of intersection of the following graphs.
   a $y = x^2 - 4x + 4$ and $y = 2x - 1$
   b $y = x^2 + 6x + 3$ and $y = 2x + 8$
   c $y = x^2 + 5x + 2$ and $y = 3x + 10$
   d $y = 2x^2 + 8x - 3$ and $y = -2x - 3$
   e $y = 3x^2 + 2x + 4$ and $y = -5x + 2$
   f $y = 5x^2 + 6x + 2$ and $y = x + 2$

Section 4: Using graphs to solve equations

HOMEWORK 8K

1 This graph represents a cyclist's training session.

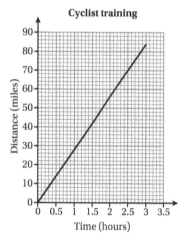

Cyclist training

a Use the graph to estimate how far the cyclist has travelled after 2 hours.
b How long did it take the cyclist to cover a distance of 42 miles?

c The equation $s = \dfrac{d}{t}$ can be used to work out the speed (s) of the cyclist.
Use values for d and t from the graph to work out the speed at which this cyclist was travelling.

2 Use this graph of the equation $y = 4x - 3$ to find the value of y for the following values:
   a $x = 0$
   b $x = 2$
   c $x = 1$

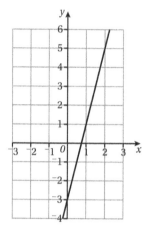

3 This graph shows how water drains from a tank at a constant rate.

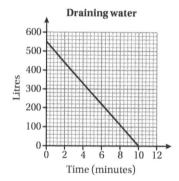

Draining water

a How much water was in the tank to start with?
b How much water was left in the tank after six minutes?
c What is the equation of the graph?

4 This diagram show the graphs of two linear equations:
$$y = 2x - 4 \quad \text{and} \quad y = {}^-3x + 1.$$

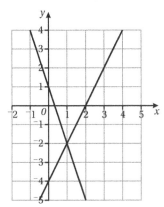

Use the graph to find the solution to the two equations.

5 This graph of a quadratic equation models the path of a ball into the air.

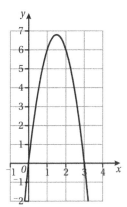

a What do you think the axes represent in this case?

b What does the y value of 0 represent?

c Are values of $y < 0$ meaningless in this context?

d Use the graph to estimate the coordinates of the maximum height of the ball.

6 Two linear graphs are shown here.

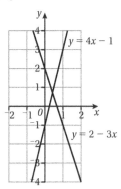

a Use the graph to estimate the values of x and y that are true for both equations.

b Find the simultaneous solution algebraically.

c What are the limitations of solving a pair of linear simultaneous equations from drawing a graph and finding the point of intersection?

7 This is the graph of a quadratic equation but the equation is not given.

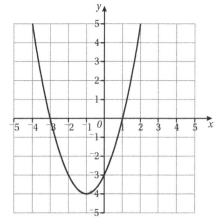

a Explain how you can use the graph to find the roots of the equation even though you don't know what the equation is.

b What are the solutions of the quadratic equation this graph represents?

c What is the quadratic equation of the graph?

d Where would the line $y = {}^-x + 2$ cut the graph? Find the solution algebraically and give your answers correct to two decimal places.

Section 5: Finding approximate solutions by iteration

HOMEWORK 8L

1 Find an approximate solution to
$x^2 - 7x + 3 = 0$
using the iteration

$$x_{n+1} = 7 - \frac{3}{x_n}$$

with $x_1 = 6$ correct to three decimal places.

2 Find an approximate solution to the square root of 21 using the iteration

$$x_{n+1} = \frac{1}{2}\left(x_n + \frac{21}{x_n}\right)$$

with $x_1 = 4$ correct to four significant figures.

3 Find an approximate solution to $x^3 - 4x + 2 = 0$ using the iteration

$$x_{n+1} = \frac{-2}{x_n^{\,2} - 4}$$

with $x_1 = 1.5$ correct to four significant figures.

HOMEWORK 8M

1 $y = x^2 - 5x - 3$

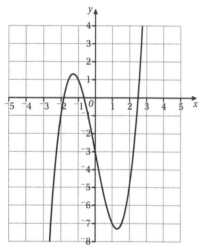

a Use the decimal search method to find the root of the cubic function
$y = x^3 - 6x - 4$
which lies between $(2, 0)$ and $(3, 0)$ correct to two decimal places.

b Use the interval bisection method to find the root of the cubic function
$y = x^3 - 6x - 4$
which lies between $(0, 0)$ and $(^-1, 0)$ correct to one decimal place.

Section 6: Using equations and graphs to solve problems

HOMEWORK 8N

1 Kimberley thinks the solutions to the equation
$x^2 + 7x + 12 = 0$ are $x = 3$, $x = 4$.
Is she right? Explain your answer.

2 Solve.
a $4x = 44$ **b** $35x - 19 = 121$
c $3(4x - 3) = 25$ **d** $3x + 12 = 5x - 4$

3 A rectangular field has a width 6 m shorter than its length and a perimeter of 120 m.
Find the length and width of the field by forming a linear equation and solving it.

4 Factorise and solve these quadratic equations.
a $x^2 + x - 12 = 0$ **b** $x^2 - 11x + 30 = 0$
c $x^2 + 4x - 12 = 0$ **d** $x^2 - 4x - 12 = 0$
e $2x^2 + 11x + 15 = 0$

5 The difference between two numbers x and y is 11.
The sum of four times x and three times y is 58.
Find the x and y.

6 A line of best fit has been drawn through this data to create a linear graph that plots height in cm against mass in kg for a specific group of people.

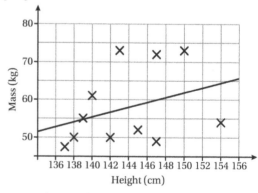

a Use the graph to estimate the mass of a person whose height is 141 cm.
b Estimate the height of a person who weighs 62 kg.

7 A father is 32 years older than his daughter.
In five years' time, he will be twice her age.
Find their present ages.

8 Solve the following equations by completing the square.

a $x^2 - 6x - 2 = 0$ b $x^2 = 7x + 6$

9 For the quadratic equation $ax^2 - 6x + 2 = 0$, find the values of a for which the equation has:

a one solution   b two solutions

c no solutions.

10 The equation $x^3 - 9x + 5 = 0$ has three solutions.

a Find an approximate solution to $x^3 - 9x + 5 = 0$ in the interval between $(0, 0)$ and $(1, 0)$ using the iteration $x_{n+1} = \dfrac{-5}{x_n - 9}$ with $x_{1 = 1.}$

Answer correct to two decimal places.

b Find a second root that lies between the interval $(2, 0)$ and $(3, 0)$ using the decimal search method.

Answer correct to two decimal places.

c Find the third root that lies between the interval $(^-3, 0)$ and $(^-4, 0)$ using the interval bisection search method.

Chapter 8 review

1 Solve for x.

a $6x - 2 = 4(2x - 3)$

b $x^2 = 15 - 2x$

c $4(x - 3) = 3(x + 12)$

d $^-x^2 = 8x + 12$

e $(x + 3)^2 = 49$

f $(x + 5)(x + 2) = 10$

2 Study this graph.

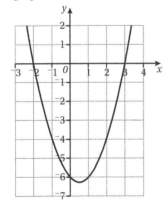

a What are the roots of the quadratic equation modelled by this graph?

b What is the equation of this graph?

3 Frank thinks that the solutions to $y = 2x + 4$ are the same as for $y = 4x + 8$.

Is he correct? Explain your answer.

4 Explain why the equations $2x + y = 6$ and $2x + y = 8$ are not simultaneous equations.

5 If you drew the graph of $y = x^2 + 2x$, where would the curve cut the x-axis?

6 The sum of two numbers is 28 and their difference is 6.

a Write a set of equations in terms of x and y to show this.

b Solve the equations simultaneously to find the two numbers.

7 Solve by the most efficient method. Leave your answers in square root form when necessary.

a $x^2 - 6x = 31$

b $x^2 + 12x = 3x - 10$

c $8 + x^2 + 14x = 5x - 12$

d $x^2 + 20x = 10$

8 An object is thrown upwards so that its height (h) in metres after a certain time (t) in seconds can be described using the formula $h = 30t - 5t^2$.

a How long does it take the object to first reach a height of 24 m?

b At what time does it come down to reach this height again?

9 The equation $x^3 - 7x + 4 = 0$ has three solutions.

a Find an approximate solution to $x^3 - 7x + 4 = 0$ in the interval between $(0, 0)$ and $(1, 0)$ using the iteration $x_{n+1} = \dfrac{-4}{x_n - 7}$ with $x_1 = 1$.

Answer correct to two decimal places.

b Find a second root that lies between the interval $(2, 0)$ and $(3, 0)$ using the decimal search method.

Answer correct to two decimal places.

c Find the third root that lies between the interval $(^-3, 0)$ and $(^-2, 0)$ using the interval bisection search method. Give your answer to two decimal places.

9 Angles

Section 1: Angle facts
HOMEWORK 9A

1. Find the value of the missing angle x in each diagram.

   a

   b

   c

2. Jenny has drawn and labelled the following diagram:

   What is wrong with Jenny's diagram?

3. What is the value of x in the diagram below?

4. In the diagram below AB and CD are straight lines intersecting at E.

   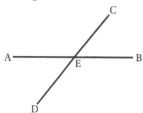

   Which angles are equal?

5. In the diagram below PQ and RS are straight lines that meet at T.
   UT is the bisector of the angle RTQ.

   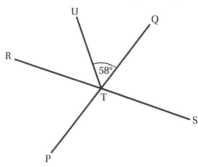

   Calculate the angles. Give reasons for your answers.

   a RTU   **b** RTS   **c** RTP
   d QTS   **e** STP

6. Five lines meet at a single point.
   The angles are given as $(x + 9)$, $(2x - 4)$, $(3x + 12)$, $(4x + 7)$ and $2x$.
   What is the value of x?

Section 2: Parallel lines and angles
HOMEWORK 9B

1. Draw and label two parallel lines and a transversal to show a pair of:
   a alternate angles
   b corresponding angles
   c co-interior angles.

Tip

When formed between parallel lines, corresponding angles are equal, alternate angles are equal and co-interior angles are supplementary (add up to 180°).

2 In the diagram below the lines AB and CD are parallel, as are AC and BD.

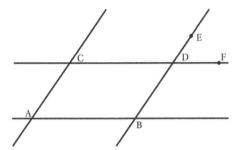

The angle CAB = 66°.

a Write down the size of angle ACD.
Give a reason for your answer.
b Write down the size of angle EDF.
Give a reason for your answer.
c Write down the size of angle ABD.
Give a reason for your answer.
d Name one other angle that is equal to CAB.
e What shape is ACDB?
Explain your answer.

3 Use the diagram below to find the following angles. Lines AF and LG are parallel, as are lines BJ and DI.
Give reasons for your answers.

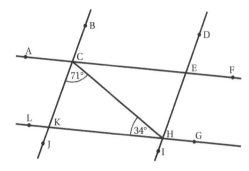

a ECH **b** ACB
c ACK **d** IHG
e CEH **f** DEF

4 In the diagram below one angle has been marked 'x'.

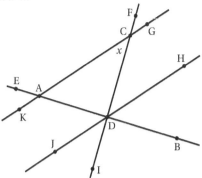

The line EB bisects the angle JDC.
Write in terms of x the values of t he angles:
a FCG **b** DCG **c** CDH **d** CDA
e JDA **f** CAD **g** KAD

Section 3: Angles in triangles
HOMEWORK 9C

1 A triangle has two angles of 54° and 74°.
What is the size of the third angle in the triangle?

2 An isosceles triangle has two equal angles of 37°.
What is the size of the third angle in the triangle?

3 The triangle ABC has an interior angle at A of 52°.
The exterior angle formed by a line through BC at C = 121°.
What is the size of the interior angle at B?

4 In the triangle DEF the interior angle at D is a right angle.

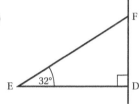

a What is the size of the exterior angle shown at F?
b If the line DE was extended beyond E, what would be the size of the exterior angle formed at E?

Tip

The exterior angle of any triangle is equal to the sum of the opposite interior angles.

5 In the diagram below the lines AB, CD and EF are parallel. The distance HJ = HG.

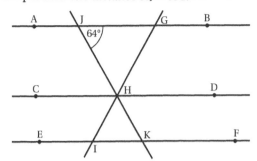

Work out, giving reasons for your answers.
a Angle HKI
b Angle HIK
c Angle IHK
d Angle JHD
e Angle GHD
f What sort of triangle is GHJ?
g Explain your reasoning.

6 a Use the diagram below and step-by-step reasoning to prove that the angles in a triangle always add up to 180°.

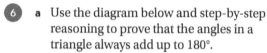

b Use the same diagram to prove that co-interior angles always add up to 180°.

Section 4: Angles in polygons

HOMEWORK 9D

1 What is the size of an interior angle at any vertex in:
a a regular pentagon?
b a regular octagon?
c a regular decagon?
d any regular polygon with n sides?

2 Calculate the sum of interior angles of a polygon with:
a 11 sides.
b 15 sides.
c 30 sides.

3 The diagram below shows a regular pentagon ABCDE.

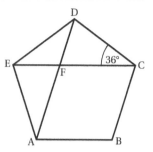

Calculate:
a The angle DFE.
b The angle EFA.
c The angle FAB.
d What shape is ABCF?
e Explain your reasoning.
f What shape is AEF?
g Explain your reasoning.

4 The regular hexagon below has been divided up into triangles.

a What type of triangles are they?
b Explain and justify your answer to part **a**.
c Write an equation for the sum of the interior angles in this diagram in terms of the number of triangles.

HOMEWORK 9E

1 What is the exterior angle at any vertex in:
a an equilateral triangle?
b a square?
c a regular hexagon?

2 Calculate the sum of the exterior angles for:
a a 12-sided shape.
b a 250-sided shape.

3 A regular polygon has the sum of the interior angles = 7740°.
a How many sides does the polygon have?
b What is the size of each interior angle?
c What is the size of each exterior angle?

Chapter 9 review

1 Use the diagram to find the following angles. Give reasons for your answers.
   a Angle EBI   **b** Angle DGF
   c Angle GFH   **d** Angle AFH

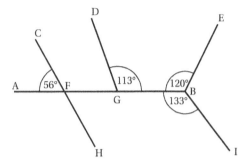

2 The diagram below shows two pairs of parallel lines.

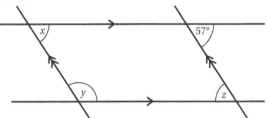

Calculate the marked angles, giving reasons for your answers.

3 One of the two equal angles in an isosceles triangle is 39°.
What are the sizes of the other two angles?

4 The triangle ABC has an exterior angle of 34° at B and an interior angle of 19° at A.
What is the size of the interior angle at C?

5 A 20-sided polygon is known as an icosagon.
   a What is the size of an interior angle of a regular icosagon?
   b What is the sum of the interior angles of an icosagon?
   c What is the size of an exterior angle in a regular icosagon?
   d What is the sum of the exterior angles in an icosagon?

6 A regular polygon has an interior angle that is $1\frac{1}{2}$ times bigger than its exterior angle.
What is the name given to this shape?

7 The exterior angle of a regular polygon is 24°.
   a How many sides does this polygon have?
   b What is the sum of the interior angles?

10 Fractions

Section 1: Equivalent fractions
HOMEWORK 10A

1 Complete each statement to make a pair of equivalent fractions.

   a $\dfrac{3}{4} = \dfrac{\square}{12}$ **b** $\dfrac{1}{3} = \dfrac{250}{\square}$ **c** $\dfrac{^-3}{^-5} = \dfrac{36}{\square}$

   d $\dfrac{\square}{24} = \dfrac{36}{16}$ **e** $\dfrac{7}{4} = \dfrac{28}{\square}$ **f** $\dfrac{20}{14} = \dfrac{50}{\square}$

2 Find the equivalent fractions of $\dfrac{3}{4}$ with:
   a denominator 32.   **b** numerator 48.
   c numerator 27.   **d** denominator 52.

3 How could you use cross multiplication to find the missing values in examples like these?

   a $\dfrac{6}{7} = \dfrac{18}{\square}$ **b** $\dfrac{\square}{54} = \dfrac{3}{18}$

4 Reduce the following fractions to their simplest form.

   a $\dfrac{3}{18}$ **b** $\dfrac{5}{20}$

   c $\dfrac{8}{10}$ **d** $\dfrac{12}{28}$

   e $\dfrac{48}{36}$ **f** $\dfrac{64}{108}$

5 Determine whether the following pairs of fractions are equivalent ($=$) or not ($\neq$).

a $\dfrac{2}{9}$ and $\dfrac{1}{4}$ b $\dfrac{2}{3}$ and $\dfrac{3}{5}$

c $\dfrac{3}{4}$ and $\dfrac{9}{15}$ d $\dfrac{20}{50}$ and $\dfrac{4}{10}$

e $\dfrac{8}{24}$ and $\dfrac{3}{9}$ f $\dfrac{12}{9}$ and $\dfrac{120}{99}$

6 Find a fraction between:

a $\dfrac{1}{4}$ and $\dfrac{3}{5}$ b $\dfrac{4}{5}$ and $\dfrac{9}{11}$

Section 2: Operations with fractions

HOMEWORK 10B

> **Tip**
>
> Learn the rules for operations with fractions – especially adding and subtracting with different denominators.

1 Work out without using a calculator. You must show all your working.

a $\dfrac{3}{5} \times \dfrac{3}{8}$ b $\dfrac{5}{11} \times \dfrac{^-5}{7}$ c $\dfrac{3}{5} \times 45$

d $\dfrac{7}{9} \times \dfrac{7}{10}$ e $2\dfrac{4}{7} \times 3\dfrac{1}{2}$ f $5\dfrac{2}{9} \times {}^-2\dfrac{1}{6}$

2 Simplify.

a $\dfrac{1}{4} \times \dfrac{3}{7} \times \dfrac{5}{9}$ b $\dfrac{2}{5} \times \dfrac{5}{8} \times \dfrac{3}{10}$

c $\dfrac{1}{3} \times \dfrac{3}{4} \times \dfrac{6}{11}$ d $\dfrac{5}{9} \times \dfrac{3}{11} \times \dfrac{9}{10}$

3 Simplify.

a $\dfrac{2}{5} + \dfrac{1}{2}$ b $\dfrac{1}{2} + \dfrac{1}{8}$

c $\dfrac{1}{4} + \dfrac{5}{8}$ d $\dfrac{5}{6} + \dfrac{3}{8}$

e $\dfrac{2}{3} + \dfrac{3}{5}$ f $\dfrac{7}{9} - \dfrac{2}{5}$

g $12 - \dfrac{1}{6}$ h $\dfrac{11}{2} - \dfrac{7}{5}$

i $2\dfrac{3}{7} + 4\dfrac{1}{3}$ j $2\dfrac{2}{5} - 1\dfrac{2}{3}$

4 Simplify.

a $\dfrac{1}{8} + \dfrac{5}{9}$ b $\dfrac{2}{11} + \dfrac{^-2}{7}$

c $\dfrac{4}{7} + \dfrac{3}{8}$ d $\dfrac{^-5}{11} + \dfrac{^-1}{3}$

e $\dfrac{3}{5} + 2\dfrac{1}{4}$ f $2\dfrac{1}{4} + \dfrac{3}{5}$

5 Calculate.

a $3 + \dfrac{2}{5} \times \dfrac{2}{5}$

b $3\dfrac{3}{4} - \left(2\dfrac{1}{4} - \dfrac{4}{15}\right)$

c $\dfrac{5}{7} \times \left(\dfrac{1}{3} + 5 \div \dfrac{2}{5}\right) + 4 \times \dfrac{2}{7}$

d $5\dfrac{7}{8} + \left(7\dfrac{1}{3} - 5\dfrac{2}{9}\right)$

6 Shamso buys a 6 kg packet of mixed nuts and raisins and she notices that $\dfrac{3}{8}$ of the contents are raisins. How many kilograms of nuts were there?

7 Josh eats 12 bananas each week. Tara eats $2\dfrac{1}{4}$ times as many.
How many bananas do they eat altogether?

8 Kalvin is a pilot.
He needs to keep track of how much time he spends in the air to maintain his flying licence. If he spends $2\dfrac{3}{4}$ hours flying on Monday, $5\dfrac{1}{5}$ hours flying on Tuesday and 75 minutes flying on Wednesday, how many hours has he spent in the air altogether?

9 $\dfrac{11}{24}$ of the people in a UK athletics team are from England, $\dfrac{1}{4}$ are from Wales, $\dfrac{1}{6}$ are from Scotland and the rest are from Northern Ireland.
  a What fraction of the team are from Northern Ireland?
  b Which country has the smallest number of team members?

10 There are $2\dfrac{1}{4}$ cakes left over after a party. These are shared out equally amongst six people. What fraction does each person get?

11 A tanker contains $56\dfrac{1}{3}$ litres of juice.
How many containers holding $\dfrac{5}{6}$ of a litre can be completely filled?

Section 3: Finding fractions of a quantity
HOMEWORK 10C

Tip

To find a fraction of a quantity you need to multiply.
Remember, any number can be expressed as a fraction with a denominator of 1.

1 Calculate.

a $\frac{5}{6}$ of 12   b $\frac{2}{9}$ of 45   c $\frac{3}{4}$ of 36

d $\frac{7}{12}$ of 144   e $\frac{4}{9}$ of 180   f $\frac{4}{5}$ of $4\frac{1}{2}$

2 Calculate the following quantities.

a $\frac{3}{4}$ of £48   b $\frac{3}{5}$ of £220   c $\frac{2}{5}$ of £45

d $\frac{3}{4}$ of $2\frac{1}{2}$ cups of sugar   e $\frac{1}{4}$ of $4\frac{2}{3}$ cakes

f $\frac{2}{3}$ of 5 hours   g $\frac{1}{3}$ of $2\frac{3}{4}$ hours

3 Express the first quantity as a fraction of the second.
a 6 p of every £1
b 25 cm of a 3 m length
c 15 mm of 30 cm
d 40 minutes per 8-hour shift
e 4 minutes per hour
f 175 m of a kilometre

4 Koresh earns £24 000 per year.
His friend Samir earns £28 000 per year.
What fraction of Samir's salary does Koresh earn?

5 The floor area of a rectangular bar is 52 m². The dance floor is 2.4 m wide and 3.6 m long. What fraction of the floor area is the dance floor?

6 A section of railway track $5\frac{3}{4}$ km long requires some maintenance work.
$\frac{3}{8}$ is completed in week 1, $\frac{3}{5}$ in week 2 and the rest is to be completed in week 3.
Calculate the length of track maintained each week.

7 A major hotel chain has 120 000 bedrooms available. On one night $\frac{1}{6}$ are sold to overseas visitors, $\frac{3}{8}$ are sold to business users, $\frac{7}{30}$ are sold to UK tourists and the rest are empty.
Calculate the quantity of rooms in each category.

HOMEWORK 10D
1 Try to write each fraction as the sum or difference of unit fractions.

a $\frac{3}{4}$ b $\frac{5}{9}$ c $\frac{7}{10}$ d $\frac{2}{5}$

2 Find three unit fractions that add up to $\frac{14}{15}$.

Chapter 10 review
1 Simplify.

a $\frac{12}{60}$ b $\frac{185}{240}$ c $4\frac{9}{36}$

2 Write each set of fractions in ascending order. Show all your working.

a $\frac{6}{7}, \frac{7}{9}, \frac{2}{3}, \frac{5}{6}$ b $\frac{14}{5}, \frac{11}{4}, 2\frac{1}{2}, 2\frac{3}{5}$

3 A rectangle has side lengths $\frac{5}{6}$ m and $\frac{2}{3}$ m.
a What is the perimeter of the rectangle?
b What is the area of the rectangle?

4 A triangle has sides of length $2\frac{1}{4}$ cm, $3\frac{2}{5}$ cm and $\frac{9}{4}$ cm.
a What is the perimeter of the triangle?
b What type of triangle is it?
The perpendicular height of the triangle is $4\frac{1}{6}$.
c What is the area of the triangle?

5 Simplify.

a $\left(\frac{5}{8} \div \frac{15}{4}\right) + \left(\frac{4}{9} \times \frac{3}{8}\right)$ b $3\frac{3}{4} \times \left(9 \div \frac{5}{8} + \frac{5}{6}\right)$

6 Lisa has $15\frac{1}{2}$ litres of water. How many bottles containing $\frac{3}{4}$ litre can she fill?

7 At a Fun Day there are 5 litres of ice cream to be sold in cones. If each cone has at least $\frac{2}{23}$ of a litre of ice cream, what is the maximum number of cones that can be made?

8 A ladder $4\frac{3}{7}$ m long is placed up against the wall.
$\frac{2}{13}$ of the length of the ladder is lost as it is placed at an angle so that it does not fall.
How high up the wall does the ladder reach?

11 Decimals

Section 1: Revision of decimals and fractions

HOMEWORK 11A

1 Write the following decimals as fractions in their simplest form.

  a 0.8   **b** 0.64

  c 2.25   **d** 0.979

  e 0.0125   **f** 0.005

  g 0.66   **h** 0.435

2 Convert the following fractions to decimals without using a calculator.

  a $\dfrac{2}{5}$ **b** $\dfrac{7}{10}$ **c** $\dfrac{11}{200}$

  d $\dfrac{3}{25}$ **e** $\dfrac{9}{20}$ **f** $\dfrac{7}{50}$

  g $\dfrac{3}{250}$ **h** $\dfrac{3}{8}$

3 Convert the following fractions to decimals using a calculator.

  a $\dfrac{1}{3}$ **b** $\dfrac{2}{9}$ **c** $\dfrac{5}{12}$

  d $\dfrac{7}{18}$ **e** $\dfrac{5}{24}$ **f** $\dfrac{1}{33}$

  g What do you notice about the denominators in each question?

  h Can you make a general rule from this?

4 Arrange the following in ascending order.

  $6\frac{3}{10}$, 6.21, $7\frac{3}{5}$, 5.98, 3.07

5 Arrange the following in ascending order.

  a 24.3, 24.72, 24.07, 24.89, 24.009

  b 0.53, 0.503, 0.524, 0.058, 0.505, 0.5

6 Fill in the boxes using <, = or > to make each statement true.

  a $\frac{1}{2}$ ☐ 0.499   **b** $\frac{2}{5}$ ☐ 0.25

  c 0.867 ☐ 0.876   **d** $\frac{5}{8}$ ☐ 0.7

  e $\frac{8}{32}$ ☐ 0.25

7 The dimensions of three cars are given in the table.

Car	Length (m)	Width (m)	Height (m)
Alfa Romeo Guilietta	4.351	1.798	1.465
BMW Z4	4.239	1.79	1.291
Jaguar F type	4.47	1.923	1.308

  a Write the cars in order of length, with the shortest first.

  b Write the cars in descending order of width.

  c Write the cars in ascending order of height.

Section 2: Calculating with decimals

HOMEWORK 11B

1 A bottle of olive oil contains 0.475 litres and costs £3.55.

  a Can you buy three bottles for £10?

  b What is the price of the olive oil per litre?

  c How many bottles would you need to buy to have at least 2 litres of olive oil?

  d At a warehouse olive oil is sold in 20 litre drums. How many bottles could you completely fill from such a drum?

  e A recipe for a salad dressing uses 15 ml. How many portions of salad dressing could be made with one bottle?

2 Estimate, then calculate. Show your working.

  a 0.7 + 0.35   **b** 13.7 − 2.9

  c 1.2 + 0.4   **d** 18.31 − 4.96

  e 3.53 × 2.4   **f** 8.99 × 5.2

3 Work out without using a calculator.

  a 13.8 + 45.6 + 3.97   **b** 34.65 + 5.08 + 2.8

  c 65.87 − 8.6   **d** 45.93 − 17.69

  e 43.9 + 9.24 − 12.16 **f** 0.87 × 100

  g 9.56 × 200   **h** 4.35 × 7.53

  i 0.564 ÷ 8   **j** 7.2 ÷ 0.8

  k 9.456 ÷ 0.4   **l** 6.84 ÷ 3.2

4 The table below shows the last six women's world records in 50 m freestyle swimming.

Holder	Time (sec)
Jill Sterkel	26.32
Kelly Asplund	26.53
Cynthia Woodhead	26.61
Anne Jardin	26.74
Johanna Malloy	26.95
Kornelia Ender	26.99

a What is the time difference between the fastest and slowest in the table?

b What is the biggest difference in the world record in the table?

c Calculate the average speed of Jill Sturkel in metres per second on her record-breaking swim. Give your answer to 1 decimal place.

5 Marita eats a 30 g bowl of porridge every morning.
She calculates that if she eats the same quantity every day for a week she will take in 14.49 g of fat, 139.16 g of carbohydrate and 35.49 g of protein.
How much of each will she eat in a single serving?

6 Sandita has £117.50 in her pocket.
She buys three tops that cost £27.99 each.
How much money will she have left?

7 The odometer in Jules' car reads 129 985.3 km when he leaves Norwich and 130 128.7 when he arrives in London. How far did he travel?

8 Jim has 2800 wooden stakes, which are each 0.6 m long.
If he laid the stakes end-to-end in a straight line, how long would the line be?

9 Juanita bought 18.9 litres of petrol at £1.37 per litre.
a How much will this cost?
b When she gets to the checkout she finds a voucher for 5 p off per litre. What is the total amount she pays?

Section 3: Converting recurring decimals to exact fractions

HOMEWORK 11C

1 State whether each of the following is a terminating decimal, recurring decimal or irrational number.

a $\sqrt{3}$ b $\sqrt{0.16}$ c 0.485

d $\dfrac{1}{7}$ e 0.66 f $\dfrac{2}{3}$

2 For the following fractions, write the denominator as a product of its prime factors, then state whether the fraction will produce a terminating or recurring decimal.

a $\dfrac{2}{9}$ b $\dfrac{7}{90}$ c $\dfrac{5}{24}$ d $\dfrac{3}{50}$

HOMEWORK 11D

1 Express each rational number as a decimal. You must show your working fully.

a $\dfrac{5}{8}$ b $\dfrac{7}{11}$ c $\dfrac{17}{9}$ d $\dfrac{17}{14}$

2 Write the following as exact fractions in their simplest form. You must show your working fully.

a $0.\dot{5}$ b $0.\dot{7}$ c $0.6\dot{4}$ d $0.7\dot{2}$

e $0.\dot{2}5\dot{3}$ f $0.\dot{7}5\dot{6}$ g $0.1\dot{3}$ h $0.25\dot{3}$

3 Work out the precise value of $0.\dot{7} + 0.\dot{1}\dot{4} + 2.\dot{6}$.

Chapter 11 review

1 Arrange each set of numbers in ascending order.

a $6.5, 6.05, 6.55, 6.501, 6.505$

b $\dfrac{2}{3}, 0.67, 0.607, 0.61, 0.66, \dfrac{5}{8}$

2 Convert to decimals and insert <, = or > to compare the fractions.

a $\dfrac{5}{8} \square \dfrac{3}{4}$ b $\dfrac{7}{10} \square \dfrac{13}{20}$

c $\dfrac{9}{13} \square \dfrac{11}{15}$ d $\dfrac{8}{9} \square \dfrac{7}{8}$

3 Write each as a fraction in its simplest terms.

a 0.46 b 0.72 c 0.08 d 0.075

4 a Increase $\dfrac{1}{5}$ by 3.4.

b Reduce 89.65 by $\dfrac{1}{4}$ of 32.8.

c Divide 6 by 0.75.   d Multiply 0.7 by 0.6.

5
a Add 46.86 and 34.08.
b Subtract 4.846 from 8.56.
c Multiply 7.84 by 200.
d Divide 29.56 by 100.
e Multiply 3.19 by 0.8.
f Simplify $\dfrac{76.8}{3.2}$.

6 Gerry and Judith have £24 each.
Gerry spends 0.425 of his money and Judith spends $\dfrac{9}{20}$ of hers.
a Who has most money left?
b How much more?

7 Convert each recurring decimal to a fraction.
a $0.0\dot{9}$ b $0.\dot{8}\dot{4}$ c $0.1\dot{6}$ d $3.4\dot{7}\dot{1}$

12 Units and measurement

Section 1: Standard units of measurement

HOMEWORK 12A

1 Convert the lengths and masses into the given units to complete the following.
a 3.6 km = ☐ m   b 45 cm = ☐ mm
c 76 m = ☐ mm   d 2.7 m = ☐ mm
e 0.04 m = ☐ cm   f 6.23 kg = ☐ g

Tip

When converting between metric units, if the unit is bigger, divide by the conversion factor. Multiply if the unit is smaller.

2 Add the following capacities.
Give your answers in the units indicated in brackets.
a 2.6 l + 6 l (ml)   b 5.1 l + 320 ml (l)
c 25 l + 3.6 l + 925 ml (l)

3 Convert each of the following into the required units.
a Total weight in kg of five bags of sultanas, each of mass 700 g.
b The length in cm of a 6.34 m long bus.
c 3472 kg of scrap metal into tonnes.

4 Grass with an area of 225 000 m² needs to be seeded. 20 g of grass seed can seed 1 square metre.
How many kg of seed are needed?

5 Convert each of the following into the required units.
a Area of 3.2 m² into mm².
b 77.46 m³ of cement into cm³.
c Engine capacity of 1295 cm³ into litres (to the nearest 0.1 litre).

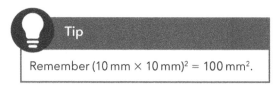

Tip

Remember (10 mm × 10 mm)² = 100 mm².

6 The diagram below shows a scale diagram of a bath.

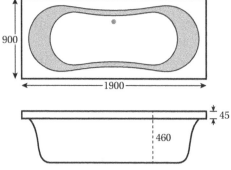

a What unit of measurement has been used?
b How high is the bath without the rim at the top?
c The plumber who installs the bath needs to know how much silicon to use to seal around the perimeter. One tube of silicon will seal 2.5 m. How many tubes will he need?

HOMEWORK 12B

1. A 10 km run starts at 10 : 45 : 00.
   The winning time is 37 minutes and 38 seconds.
   a At what time does the winner cross the line?
   b The runner who comes second crosses the line in 38 minutes and 5 seconds. How far behind the winner was she?
   c The third placed runner finishes a further 36 seconds behind. At what time does she cross the finishing line?

2. Hannah has a baby at 2.35 pm on 8 March 2013. Calculate her baby's age at the same time on 14 February 2014 in:
   a weeks. b days. c hours. d seconds.

3. The table gives the value of the pound against four other currencies in August 2014.

British pound (£)	1
Euro (€)	1.25
US dollar ($)	1.67
Australian dollar (AS$)	1.79
Indian Rupee (INR)	101.88

   a Calculate the value of each of the other currencies in pounds at this rate (to the nearest penny).
   b Convert £175 to US dollars.
   c How many Indian rupees would you get if you converted £65 at this rate?
   d Dilshaad has 9000 Indian rupees. What is this worth in pounds at this rate?

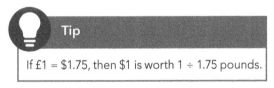

Tip

If £1 = $1.75, then $1 is worth 1 ÷ 1.75 pounds.

Section 2: Compound units of measurement

HOMEWORK 12C

Tip

Compound measures are rates, like miles per hour or metres per second.

1. Henri earns £7.85 per hour.
   One week he worked 37.5 hours.
   Sian earned £196.35 for working 21 hours.
   How much more does Sian earn per hour?

2. A plasterer can plaster an area of 45 m² in $4\frac{1}{2}$ hours.
   On average, what area can she plaster in 15 minutes?

3. Jake runs 42 km in 3 hours and 12 minutes. What is his average speed?

4. A car travels 504 km at an average speed of 96 km/h. How long does this journey take?

5. Paralympian Jonnie Peacock won the 100 m at the London Para-Olympics in 2012 with a time of 10.9 seconds.
   a Express this speed in m/s. Give your answer to three significant figures.
   b How fast is this in kilometres per hour? Give your answer to the nearest whole number.

Tip

Think about how many metres make a kilometre and how many seconds make an hour.

6. An aircraft takes off from Alicante at 11.20 a.m. (European time) and lands at Stansted Airport at 1:40 p.m. UK time. European time is one hour ahead of UK time, and Alicante is 1509 km from Stansted.
   Calculate the average speed of the aircraft.

7. An Ironman Triathlon is a 2.4 mile swim, 112 mile cycle ride and 26.2 mile marathon.
   The 2013 world championship was won in 8 : 12 : 29 (men) and 8 : 52 : 14 (women).
   Calculate the difference in the average speed of the men's and women's champions.

HOMEWORK 12D

1. A cuboid of material with side lengths 10, 20 and 30 cm has a mass of 0.534 kg.
   Calculate the density of the material in g/cm³.

2. Calculate the volume (in cm³) of a piece of wood with a mass of 0.275 kg and a density of 0.9 g/cm³.

3 A bus exerts a force of 8600 N on the road, spread evenly over its six tyres.
Each of the six tyres has an area of 0.15 m² in contact with the road.
What pressure does the bus exert on the road through each tyre?

4 A metal block has a weight of 22 N.
The block is a cuboid with sides 100, 150 and 175 cm long.
Calculate the pressure exerted by the block in N/m² when each of the sides is in contact with the floor.

Section 3: Maps, scale drawings and bearings

HOMEWORK 12E

Tip

Scales are a way of representing big distances on smaller maps and diagrams. A scale of 1 : 100 means 1 unit is being used to represent 100 units in reality.

1 Work out the real distance (in kilometres) that a map distance of 72 mm would represent at each scale.
a 1 : 100   **b** 1 : 1500   **c** 1 : 18 000
d 1 : 100 000   **e** 1 : 2 000 000

2 Arshwin says that a map drawn to a scale of 1 : 20 000 is a larger scale map than one drawn to 1 : 200 000.
Is he correct?
Explain your answer

3 A flight from Norwich to Manchester takes 45 minutes.
On a map with a scale of 1 : 1 000 000 the distance from Norwich to Manchester is 30 cm.
a Calculate the distance flown in kilometres.
b What was the plane's average speed on this flight?

4 Toy cars are manufactured using a scale of 1 : 43.
Work out:
a the height of a car if the model is 3.2 cm high
b the length of a car if the model is 9.7 cm long
c the length of the model if the real length is 4.5 m.

5 A SatNav has a variable scale. At one point it is showing a journey of 165 km as a distance of 4.125 cm on the screen.
a What is the scale of the map at this point?
b The driver only wants to know what turnings there are in the next 1 km. If the screen is a 5 cm square, suggest a good scale that will show the next km.

HOMEWORK 12F

1 A garden in a new house is 55 m long and 12 m wide.
Draw scaled diagrams to show what it would look like at each of these scales.
a 1 : 20   **b** 1 : 50   **c** 1 : 75

2 An architect is drawing a plan of a house to a scale of 1 : 50.
a What should the scaled dimensions of the kitchen be if the real dimensions are 5900 mm by 3600 mm?
b Calculate the scaled length of the kitchen island if it is 1.6 metres long in reality.

HOMEWORK 12G

1 Write the three-figure bearing that corresponds with each direction.
a Due east
b South-west
c North-west

2 Use a protractor to measure the bearing of each plane from base on the diagram.

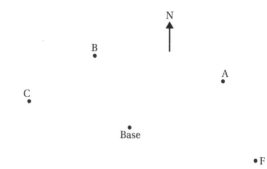

3 Village A is 75 km from village B on a bearing of 105°.
Village C is 8 km from village A on a bearing of 235°.
Use a suitable scale and draw a scale diagram and find:
  a the distance from B to C.
  b the bearing of B from C.

Chapter 12 review

1 Work out:
  a The number of seconds in four days.
  b The number of kilometres travelled in $5\frac{1}{2}$ hours by a car travelling at 67 km/h.
  c The distance in km in real life of a length of 7.5 cm on a map with a scale of 1 : 20 000.
  d The number of litres in 525 000 millilitres.
  e The capacity of a fish tank that is 50 cm × 25 cm × 40 cm, in litres.

2 The distance between London and Edinburgh is 535 km.
How far would this be on a map scale of 1 : 2 000 000?

3 Convert 75 000 cm² into m².

4 How many mm² are there in 5 m²?

5 A cycle is travelling at an average speed of 30 km/hour for two hours on a bearing of 075°.
Use a scale of 1 cm to 10 km to show this journey.

6 The density of an object is 12 kg/m³.
Work out the mass of 35 m³ of the object.

7 A yacht sails due east from point A for 20 km to reach a buoy at point B.
She then travels 12 km on a bearing of 225° from B to reach a buoy at point C.
  a Use a scale of 1 cm to 2 km to represent her journey on a scale diagram.
  b Find the bearing from C to A.
  c Find the direct distance from C to A in kilometres.
  d If it took the sailor $1\frac{1}{4}$ hours to sail back directly from C to A, find the yacht's average speed:
    i in km/h.   **ii** in m/s.

13 Percentages

Section 1: Review of percentages
HOMEWORK 13A

1 Write each of the following as a percentage.
  a 0.75   **b** 0.6   **c** 0.08   **d** 0.632
  e $\frac{1}{2}$ **f** $\frac{4}{5}$ **g** $\frac{7}{8}$ **h** $\frac{1}{20}$

Tip

'Percent' means 'out of a hundred'.

2 Write each of following percentages as a common fraction in its simplest terms.
  a 45%   **b** 90%   **c** 70%
  d 37.5%   **e** 40%   **f** 88%
  g 65%   **h** 32%

3 Write the decimal equivalent of each percentage.
  a 78%   **b** 99%   **c** 35%
  d 48%   **e** 0.6%   **f** 0.09%

4 **a** If 94.5% of houses in the UK have a TV set, what percentage do not?
  b If $\frac{3}{4}$ of mobile phones are sold as pay-as-you-go, what percentage are not?
  c 0.245 of computer users back up their work every day. What percentage do not do this?

5 Bjorn spends 31.7% of a day asleep, 0.127 of the day doing housework and $\frac{4}{9}$ of the day at work.
What percentage of the day is spent doing other things?

6 Anya pays 0.12 of her salary into her savings account.
What percentage of her salary is this?

7 Barry gets the following marks for three tests: $\frac{31}{50}, \frac{17}{35}, \frac{51}{80}$.

a In which test did he get the best marks?

b What is his mean result for the three tests as a percentage to the nearest whole number?

Section 2: Percentage calculations
HOMEWORK 13B

>
> **Tip**
>
> To find a percentage of a quantity, multiply by the percentage and divide the result by 100.

You are allowed to use a calculator for these questions.

1 Calculate.

a 5% of 150 **b** 8% of 300 **c** 30% of 150

d 22% of 50 **e** 120% of 70 **f** 150% of 80

2 Calculate, give answers as mixed numbers or decimals as necessary.

a 17% of £40 **b** 70% of 65 kg

c 3.5% of 80 minutes **d** 6.3% of £1000

e 4.7% of 210 m

3 Annie got 65% for a test that was out of 60 marks.

a What is her mark as a percentage?

She gets the same percentage score in a test that is out of 80.

b How many marks did she get in this test?

4 A machine in a factory has a failure rate of 2.5%. If the machine makes 15 200 items per day, how many are defective?

5 Approximately 72% of homes put the correct recycling bin out every week, the rest do not. In a city of 32 904, how many homes:

a do put out the correct bin?

b don't put out the correct bin?

6 A laptop is advertised for sale for £549 excluding VAT.

VAT is charged at 20%.

a How much extra is the VAT?

b How much is the total cost of the laptop?

7 9.5% of a 864 hectare farm grows wheat and the rest is used to grow barley.

How many hectares of land are used to grow:

a wheat? **b** barley?

8 The cost of energy used in Banjul's house increases by 63% in the three months of winter compared to the three months of summer. If the cost in the summer is £205:

a what is the increase in cost in the winter?

b how much is the fuel bill for the winter?

>
> **Tip**
>
> To find one quantity as the percentage of another, divide the first quantity by the second and multiply the result by 100.

9 The number of hits on a variety of websites are given in terms of their percentage change.

Website	Number of hits 2013	Number of hits 2014	% change in hits
MyTube	24 581 548	25 125 495	2.2%
Cheeper	5 248 559	8 458 125	61.2%
HiSpace	57 489 326	56 157 489	2.3%
FileShare	214 548	584 725	172.5%

a Which website can claim to be the fastest growing?

b Which website looks to be losing popularity?

c Which website would you rather be the owner of? Explain your answer.

10 If A is 75% of B, what percentage of B is A?

HOMEWORK 13C

1 Express the first amount as a percentage of the second.

Give your answer correct to no more than two decimal places (if necessary).

a 200 m of 4 km **b** 32 m of 4 km

c 125 m of 2 km **d** 12 cm of 3 m

e 15 mm of 6 cm **f** 23 cm of 5 m

g 45 p of £6 **h** 62 p of £4.50

i 12 seconds of a minute

2 Sanjita got 21 out of 26 for an assignment and Niall got 23 out of 30.

Who got the highest percentage mark? What was this mark?

3 In a school election 1248 students were eligible to vote.
Of these, 1008 voted.
What was the percentage voter turnout?

4 Mel improved his swimming time for the 200 m backstroke race by 3 seconds.
If his previous best time was 2 minutes and 20 seconds, what is his percentage improvement?
Give your answer to one decimal place.

Section 3: Percentage change
HOMEWORK 13D

1 Increase each amount by the percentage given.
a £36 by 20% b £600 by 45%
c £40 by 6.5% d £4500 by 12%
e £625 by 4% f £456 by 4.6%

Tip

The quick way to find a percentage increase or decrease is to find the single multiplier.

2 Decrease each amount by the percentage given.
a £54 by 10% b £560 by 24%
c £862 by 14.5% d £632 by 7.4%
e £278 by 5.3% f £34 900 by 15.7%

3 House prices in a city have increased by 3.1% in a year.
How much would a house be worth at the end of the year if it was priced at £345 000 at the start of the year?

4 Sammy has been offered a pay rise of £20 per week or a 5% increase on his hourly rate.
If he earns £7.40 per hour for a 37.5 hour week, would he be better taking the £20 or the percentage increase? How much will he now earn?

5 The price of being a member of a golf club has decreased by 15% in a year.
If last year's price was £485, what is the price this year to the nearest £?

6 Shares in a building company are worth £12.15 each initially. After six months their value has decreased by 20%.
a Find their value after six months.
b Six months later their value has risen again by 20%.
Are the shares worth the same amount as they were originally?
Explain your answer

7 New cars depreciate in value (their value decreases) significantly in their first year.
Copy and complete the table to show the value of these cars at the end of their first year:

Car	Original price	Depreciation	New value
Car A	£26 200	32%	
Car B	£16 800	17%	
Car C	£36 500	26%	

8 In a small village school nine pupils are in Year 6. This is 15% of the school population.
a How many pupils are there in the school?
b How many pupils are not in Year 6?

9 The value of a company drops by 14% on the stock market.
The following day the original value is restored.
What was the percentage increase from the first day to the second day?

10 Sybil invests £250 at 5% compound interest.
a How much money does she have after five years?
b How many years does she need to invest the money before it is greater than £500?

11 A new car is sold for £29 500. It depreciates by 21% in each of its first three years.
a How much is it worth after three years?
The car then depreciates at a rate of 12% per year.
b How old will the car be before it is worth less than £10 000?

HOMEWORK 13E

1 Find the original values if:
a 20% is £8 b 6% is 39.6 kg
c 140% is 840 g d 105% is £472.50

2 Mick has had a 7.4% pay rise.
If his new salary is £36 730.80, what was his salary before the rise?

Tip

The quick way to find a percentage increase or decrease is to find the single multiplier.

3 Keith is training for a triathlon. His overall time has improved by 3% since he began his training. His new personal best is now 5 hours and 24 minutes. What was his old personal best (to the nearest minute)?

Chapter 13 review

1 Write as fractions in their simplest form.
  a 35%   **b** 20%   **c** 2.5%

2 Express each of these fractions or decimals as a percentage.
  a $\dfrac{4}{5}$ **b** $\dfrac{1}{3}$ **c** $\dfrac{7}{16}$
  d 0.6   **e** 0.07   **f** 0.003

3 The value of a vintage car increased from £140 000 to £151 200.
What percentage increase is this?

4 The value of a car was £45 600 when it was new. After two years it has lost 54.5% of its value.
How much is it worth after two years?

5 Vlad has a machine that makes golf tees at 45 per minute.
He wants to increase the speed of the machine by 9%.
How many whole tees will the machine make per minute if he is successful?

6 Express.
  a 45 p as a percentage of £3.
  b 240 g as a percentage of 4 kg.

7 The table below shows the ten most popular marathon races in 2013.

Marathon	Number of starters	Number of finishers
New York	47 000	46 759
Chicago	45 000	37 455
Berlin	40 987	34 377
Paris	40 000	32 980
London	37 000	36 672
Tokyo	36 000	34 678
Osaka	30 000	27 157
Honolulu	30 000	23 786
Marine Corps (Washington D.C.)	30 000	23 515

  a Which of these races can claim to have the best finishing percentage?
  b Express the number of starters in London as a percentage of the number of starters in New York.
  c Can you use the data to explain which marathon is the toughest?

8 The average (mean) house price in the UK at the end of 2013 was £250 000 (source ONS).
  a If house prices rise by 8% on average in 2014, what would be the average house price at the end of 2014?
  b House prices increased by 5.5% in 2013. What was the average price at the end of 2012?

9 Keith is training for a triathlon.
His overall time has improved by 3% since he began his training.
His new personal best is now 5 hours and 24 minutes.
What was his old personal best (to the nearest minute)?

10 A market trader normally makes a profit of 25% on sales of batteries on his stall.
Because of increased competition he reduces the selling price of his batteries by 10%.
What is the cost price of a pack of batteries that sells for £5.40?

14 Algebraic formulae

Section 1: Writing formulae

HOMEWORK 14A

1. A photocopy shop charges a £2.50 service fee and £0.12 per page for bulk photocopying.
   a. Write a formula for the total cost £C of a photocopying job with n pages.
   b. The shop decides to increase the service fee to £2.80 and drop the cost per page by 2p per page. Write a revised formula for finding the total cost.

> **Tip**
>
> When you write a formula always say what the variables represent.

2. A stand for a display is made from two circular metal rings joined by four upright metal rods. The top ring has radius q cm, the bottom has radius p cm and the rods are h cm long. Write a formula for finding the total amount of metal (T) in terms of p, q and h.

3. VAT of 20% is added to the price of goods. Let P be the price excluding VAT and I be the price including VAT. Write a formula that shows the relationship between P and I.

4. Three numbers are represented by x, $3x$ and $x + 2$.
   a. Write formula for finding the sum (S) of the three numbers.
   b. Write a formula for finding the mean (M) of the three numbers.

5. Pesticide is applied to a crop at a rate of 1.25 kg per hectare. Write a formula for finding how much pesticide (P) you need to use for a rectangular field L m long and B m wide.

> **Tip**
>
> 1 hectare = 10 000 m^2.

Section 2: Substituting values into formulae

HOMEWORK 14B

1. a. $M = 9ab$ — Find M when $a = 7$ and $b = 10$.
   b. $V - rs = 2uw$ — Find V when $r = 8$, $s = 4$, $u = 6$ and $w = 1$.
   c. $\dfrac{V}{30} = h$ — Find V when $h = 25$.
   d. $P - y = x^2$ — Find P when $x = 2$ and $y = 8$.

2. Substitute the given values into each formula and calculate the unknown value. If necessary, round answers to four significant figures.
   a. $E = \dfrac{1}{2}mv^2$ — $m = 10$, $v = 2.5$
   b. $v = u + at$ — $u = 7$, $v = 12$ and $t = 5$
   c. $D = \dfrac{a}{(1 - r)}$ — $D = 8$ and $a = 4$
   d. $v = \dfrac{5wh}{6n}$ — $w = 2.8$, $h = 36$ and $n = 8$
   e. $r = \dfrac{1}{2}(56 - t)$ — $t = 14.75$
   f. $px - q = rx + t$ — $p = 8$, $q = {}^-4$, $r = 3$ and $t = 7$

3. The pressure (P) of a given amount of gas is inversely proportional to the volume (V) of the gas. The relationship between P and V can be expressed as the formula:
   $$PV = k$$
   Given that P is 60 when V is 40, find the value of P when V is 30.

Section 3: Changing the subject of a formula

HOMEWORK 14C

1. Make m the subject of $D = km$.

2. Make c the subject of $y = mx + c$.

3. Given that $P = ab - c$, make b the subject of the formula.

4 Given that $a = bx + c$, make b the subject of the formula.

5 Change the subject of each formula to make the variable in the bracket the subject.

a $U + T = V + W$ (V)
b $3(V + W) = U - T^2$ (V)
c $A = \dfrac{C}{B}$ (B)
d $A = \left(\dfrac{B}{C}\right)$ (B)
e $2P = Q^2$ (Q)
f $P = 2Q^2$ (Q)
g $P = Q^2R$ (Q)
h $Q = \sqrt{(2P)}$ (P)
i $Q = \sqrt{(PR)}$ (P)
j $Q = \sqrt{(P - R)}$ (P)
k $PQ = R^2$ (Q)

6 Ohm's Law is a formula that links voltage (V), current (I) and resistance (R). Given that

$$V = IR:$$

a change the subject of the formula to I.
b find the current (in amps) for a voltage of 50 volts and a resistance of 2.5 Ohms.

7 The area of a circle can be found using the formula $A = \pi r^2$.

a Change the subject of the formula to r.
b Find the length of the radius of a circle with an area of 100 mm². Give your answer correct to 3 sf.

8 Rearrange these formulae to make x the subject.

a $ax - b = cx + d$ b $\dfrac{q}{x} + p = r$

c $z = \dfrac{(x + y)}{(x - y)}$

Section 4: Working with formulae
HOMEWORK 14D

1 The perimeter of a rectangle can be given as $P = 2(l + b)$, where P is the perimeter, l is the length and b is the breadth.

a Make b the subject of the formula.
b Find b if the rectangle has a length of 45 cm and a perimeter of 161 cm.

Tip

When you are given shape problems it is helpful to draw a diagram and label it to show what the parts of the formula represent.

2 The circumference of a circle can be found using the formula $C = 2\pi r$, where r is the radius of the circle.

a Make r the subject of the formula.
b Find the radius of a circle of circumference 56.52 cm. Use $\pi = 3.14$. Give your answer to one decimal place.
c Find the diameter of a circle of circumference 144.44 cm. Use $\pi = 3.14$. Give your answer to one decimal place.

Tip

When you are given a value for π you must use the given value to get full marks. Using exact values can lead to rounding differences.

3 To convert temperatures from Celsius to Fahrenheit you can use the formula $F = 1.8C + 32$, where F is the temperature in degrees Fahrenheit (°F) and C is the temperature in degrees Celsius (°C).

a Rearrange the formula to convert from Fahrenheit to Celsius.
b Convert:
  a 15°C to °F
  b 92°F to °C
  c 12°F to °C
  d ‾6°C to °F
c Normal adult body temperature is about 37°C. What is this in °F?
d Water boils at 100°C at sea level. What is the Fahrenheit equivalent of that temperature?
e The lowest temperature ever recorded on Earth was ‾129°F (in Antarctica). What is this in °C?
f Temperature can also be measured in units called Kelvin. Find out how this scale works and explain how to convert temperatures in °C to Kelvin.

4 An airline uses the formula $T = 70P + 12B$ to roughly estimate the total mass of passengers and checked bags per flight in kilograms. T is the total mass, P is the number of passengers and B is the number of bags.

a What mass does the airline assume for:
  i a passenger?   ii a checked bag?
b Estimate the total mass for 124 passengers, each with two checked bags.
c Make B the subject of the formula.
d Calculate the number and the total mass of the bags if the total mass of 124 passengers and their checked bags on a flight is 9.64 tons.

5 When an object is dropped from a height, the distance (m) in metres that it has fallen can be related to the time it takes for it to fall (t) (in seconds) by the formula $m = 5t^2$.

a Make t the subject of the formula.
b Calculate the time it takes for an object to fall from a distance of $180\,\text{m}$.

6 Pharmacists use special formulae to calculate correct dosages of medicines for children. Two common formulae are given here:

Young's rule: $\left(\dfrac{\text{child's age in years}}{\text{child's age in years} + 12}\right) \times$ adult dose

Fried's rule: $\left(\dfrac{\text{child's age in months}}{150}\right) \times$ adult dose.

a Rewrite both rules using D for child's dose, m for age in months, y for age in years and a for adult dose.
b An adult dose of a particular medicine is 65 mg.
Calculate the correct dosage for a $3\frac{1}{2}$-year-old child using both formulae.
Comment on any differences.

Chapter 14 review

1 Change the subject of each formula to the letter given in square brackets.

a $x + \dfrac{1}{y} = z$ $\qquad [y]$

b $a + \dfrac{1}{\sqrt{b}} = c$ $\qquad [b]$

c $\sqrt{x} + \dfrac{1}{\sqrt{x}} = \dfrac{1}{\sqrt{x}\sqrt{y}}$ $\qquad [x]$

d $\sqrt{x} + \dfrac{1}{\sqrt{x}} = \dfrac{1}{\sqrt{x}\sqrt{y}}$ $\qquad [y]$

e $xy + x = z$ $\qquad [x]$

f $yz + y = xy + z$ $\qquad [y]$

g $p = q + \dfrac{1}{q} - 1$ $\qquad [q]$

h $y = \dfrac{\sqrt{x} + z}{\sqrt{x} - z}$ $\qquad [z]$

i $y = \dfrac{\sqrt{x} + z}{\sqrt{x} - z}$ $\qquad [x]$

2 Given $I = \dfrac{E}{R}$, find I when $E = 250$ and $R = 125$.

3 $P = \dfrac{t - m}{d}$

If $P = 12$, $t = 16$ and $m = 8$, what is d?

4 a $y = \dfrac{12}{x} + 2$

Find y when $x = {}^-3$.

b $y = (x + 3)(x - 1)$.
Find x when $y = {}^-3$.

c $y = 2^x$
Find y when $x = {}^-4$.

5 The formula $d = \dfrac{v}{10} + 2$ can be used to work out how many car lengths you should leave between you and the car in front of you when you are driving at v km/h.

a How many car lengths should you leave between you and the car in front of you when you are travelling at 100 km/h?
b If you are obeying this rule and you have left 7.5 car lengths space in front of you, what is your speed?

6 The height of a golf ball at any moment during its flight can be found using $h = {}^-4t^2 + 24t$, where h is the height above the ground in metres and t is the time in seconds.

a What is the height of the ball after 3 seconds?
b Find t when h is 20 m.

15 Perimeter

Section 1: Perimeter of simple and composite shapes

HOMEWORK 15A

1 Calculate the perimeter of each shape.

a
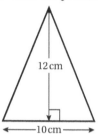

b
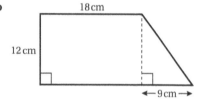

c
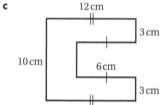

d
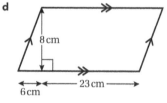

2 Find the perimeter of each shape.

a
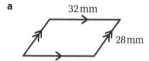

b 11.25 cm
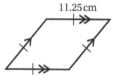

c 19 mm 45 mm
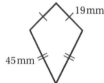

d 21 mm 14 mm
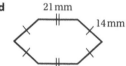

e 1.5 cm 5.3 cm 6.8 cm 3.4 cm 4.9 cm
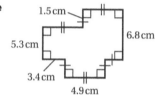

f 92 mm 7.2 cm 69 mm
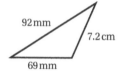

3 Determine the perimeter of each shape.
   a An equilateral triangle with sides of length 12.6 cm.
   b A square with sides of length $2x$ cm.
   c A rectangle which is 132 mm long and 6.5 cm wide.

4 Find the cost of fencing a triangular plot 45 m long and 37 m wide if the fencing costs £23.80 per metre.
Leave 2.5 m unfenced for the gate.

5 A rectangular field has a perimeter of 326 m. The width is at least 60 m and the length is at least 75 m.
Suggest five possible sets of dimensions for the field.

6 The ground set aside for allotments in a particular village is a rectangular field with side lengths of 40 m and 36 m.
The community plans to plant apple trees around the boundaries of the field.
Each tree needs to be planted at least 10.5 m apart.
a What is the perimeter of the field?
b What is the maximum number of apple trees that can be planted around the boundaries of the field?

7 Six square tiles are arranged in a row. The perimeter of the rectangle formed by the tiles is 315 cm.
a What is the length of the sides of the square tiles?
b If the same tiles were rearranged as opposite, what would the perimeter of the rectangle be?

HOMEWORK 15B

1 The perimeter of each shape is given. Find the value of x.

a
12 cm
x
$P = 34$ cm

b 4x
x
$P = 125$ cm

c
x
$P = 26$ cm

d 43 mm
x
$P = 128$ mm

e x
$P = 51$ mm

f x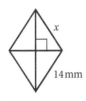
14 mm
$P = 42$ mm

g 2 mm + 2x
2x
$P = 520$ mm

h 2x
x x
$\frac{1}{2}x$ $\frac{1}{2}x$
x x
x
$P = 180$ cm

2 Each of these shapes has a perimeter of 30 cm. Determine the length of the unknown side/s in each shape.

a 8 cm x
11 cm

b x

c 11 cm
x

d
12 cm
x

e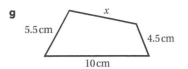
10 cm
x

f 6 cm
y

g x
5.5 cm
4.5 cm
10 cm

h 9 cm
5 cm
x

3 An isosceles triangle has a perimeter of 28 cm. The equal sides are x m long and the third side is 100 mm long. What is the length of each equal side?

Tip

Make sure all measurements are in the same units before you do any calculations.

4 A rectangle is three times as long as it is wide. If it has a perimeter of 480 mm, what are the dimensions of the sides?

5 A rhombus has a perimeter of 1.2 m. Another, smaller rhombus is drawn inside it, with its diagonals intersecting in the same place. It has a perimeter of 80 cm. What is the perpendicular distance between the two rhombuses?

Section 2: Circumference of a circle

HOMEWORK 15C

Give all your answers correct to three significant figures.

1. Find the perimeter of each of these shapes.

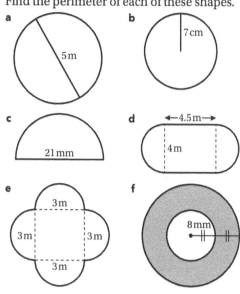

a

5 m

b

7 cm

c

21 mm

d

←4.5 m→

4 m

e

3 m

3 m

3 m

3 m

f

8 mm

2. Calculate the circumference of a circle with diameter:
   a 21 m.   b 91 mm.   c 3.4 cm.   d 4.08 cm.

3. The rim of a bicycle wheel has a radius of 31.5 cm.
   a What is the circumference of the rim?
   b The tyre that goes onto the rim is 3.5 cm thick.
     Calculate the circumference of the wheel when the tyre is fitted to it.

4. How much string would you need to form a circular loop with a diameter of 28 cm?

5. What is the radius of a circle of circumference:
   a 14 mm?   b 81 cm?
   c 206 mm?   d 31.5 cm?

6. The distance around the outside edge of a dartboard is 102 cm. Determine the distance from the centre of the bulls-eye in the middle of the dartboard to the edge of the board.

HOMEWORK 15D

1. Find the perimeter of each shape.
   In your calculations, use the value of π given by your calculator.
   Give your answers correct to three significant figures.

a

60° 8 cm

b

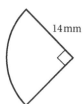

14 mm

c

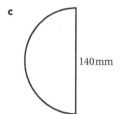

140 mm

d

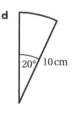

20° 10 cm

e

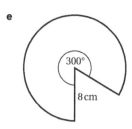

300°

8 cm

2. Determine the length of the marked arc in each circle.

a

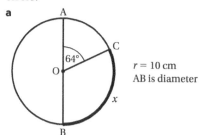

A

C

64°

O

x

B

$r = 10$ cm
AB is diameter

b

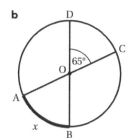

D

C

65°

O

A

x

B

$r = 6$ cm
AC = BD = diameter

c

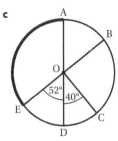

AD is diameter
$r = 5\,cm$

d

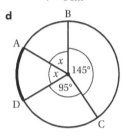

$r = 12\,cm$

③ Nicky is stuck in traffic. She has driven 23 m from point A to point B around a traffic circle of radius 14 m.

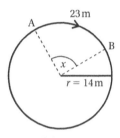

a Determine the circumference of the traffic circle.
b What is the size of the angle at x?

④ Calculate the length of the arc AB, subtended by the given angle, in each figure.

a

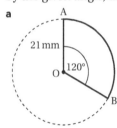

b

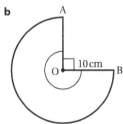

⑤ The diagram shows a cross-section of the Earth.

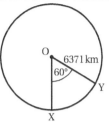

Two cities, X and Y, lie on the same longitude. Given that the radius of the Earth is 6371 km, calculate the distance, XY between the two cities.

Section 3: Problems involving perimeter and circumference
HOMEWORK 15E

① Determine which shaded shape has the greater perimeter and say how much greater it is.

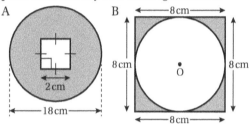

② This diagram shows a silver earring.

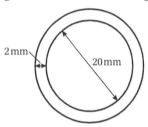

Calculate, correct to one decimal place, the circumference of:
a the inner circle of the earring
b the outer circle of the earring.

③ Find the distance around this track.

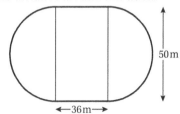

47

4 Look at this diagram of a dartboard carefully.
The dartboard has a diameter of 41 cm.
The divisions between numbers and sections
are outlined in thin metal wire (indicated by
the black lines on this diagram.)
Use the given dimensions to work out the total
length of wire on this board. Show all your
calculations.

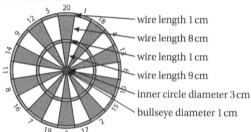

wire length 1 cm
wire length 8 cm
wire length 1 cm
wire length 9 cm
inner circle diameter 3 cm
bullseye diameter 1 cm

Chapter 15 review

1 Calculate the perimeter of each shape.

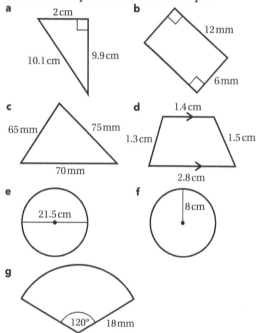

a 2 cm, 10.1 cm, 9.9 cm

b 12 mm, 6 mm

c 65 mm, 75 mm, 70 mm

d 1.4 cm, 1.3 cm, 1.5 cm, 2.8 cm

e 21.5 cm

f 8 cm

g 120°, 18 mm

2 Determine the perimeter of each of these shapes.

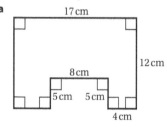

a 17 cm, 12 cm, 8 cm, 5 cm, 5 cm, 4 cm

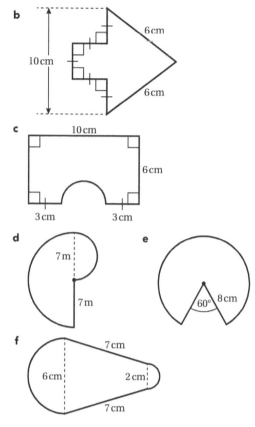

b 10 cm, 6 cm, 6 cm

c 10 cm, 6 cm, 3 cm, 3 cm

d 7 m, 7 m

e 60°, 8 cm

f 7 cm, 6 cm, 2 cm, 7 cm

3 A pizza company advertises the following
pizza sizes.

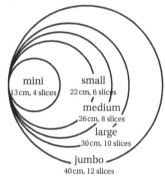

mini
13 cm, 4 slices

small
22 cm, 6 slices

medium
26 cm, 8 slices

large
30 cm, 10 slices

jumbo
40 cm, 12 slices

a Calculate the circumference of each pizza
size offered.

b The pizzas are cut into different numbers of
slices. Assuming each slice of a particular
pizza is the same size, draw a diagram to
show the dimensions of one slice of each
pizza size. Include the arc lengths and the
size of the angle at the centre.

c What is the length and width of the smallest
square box a slice of the jumbo pizza can fit
in?

16 Area

Section 1: Area of polygons
HOMEWORK 16A

1 Write a formula for finding the area of each shape.

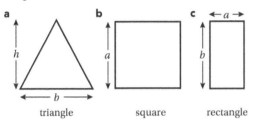

triangle   square   rectangle

2 Determine the area of each shape. All lengths are given in centimetres.

a
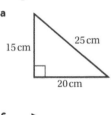
15 cm   25 cm   20 cm

b

7 cm   12 cm

c
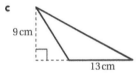
9 cm   13 cm

d
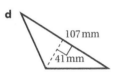
107 mm   41 mm

e
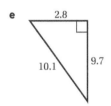
2.8   10.1   9.7

f
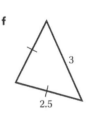
3   2.5

g
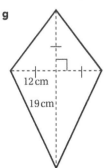
12 cm   19 cm

h
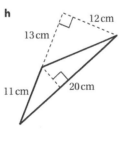
12 cm   13 cm   11 cm   20 cm

i
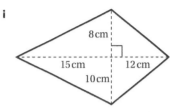
8 cm   15 cm   12 cm   10 cm

3 A triangle has a base of 9 cm and a height of 5 cm. What is its area?

4 A triangle of area 80 cm² has a base of length 15 cm.
What is its height?

5 A triangular sail has a height of 5.65 m and an area of 68.93 m².
What is the length of its base?

6 These two symmetrical arrow designs are used as markings at a sports ground.

Arrow A
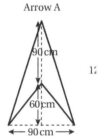
90 cm   1?   60 cm   90 cm

Arrow B
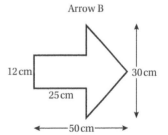
12 cm   25 cm   30 cm   50 cm

Determine the area of each arrow.

> **Tip**
>
> You learned how to convert units of area in Chapter 12 of the Student Book. Read through that again if you have forgotten how to do this.

7 The arrows from question 6 are painted white. One tin of white paint covers an area of 0.8 m². Work out how much paint you would need to paint:

a 35 arrow A designs

b 20 arrow B designs

c 100 arrow A designs and 125 arrow B designs.

HOMEWORK 16B

1 Write a formula for determining the area of each shape.

a

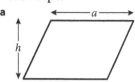

parallelogram

b

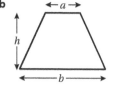

trapezium

c

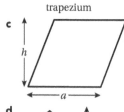

d

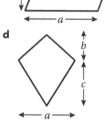

2 Determine the area of each shape.

a

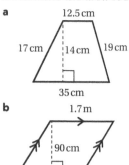

b

c

d

e

f

g

h

i

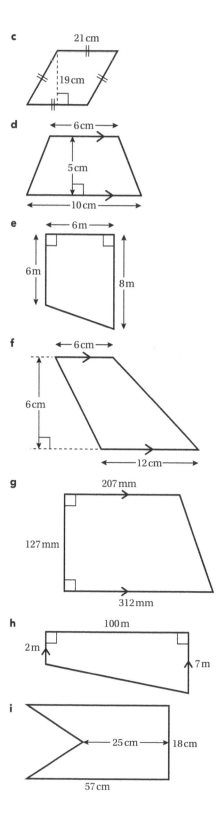

3 Calculate the area of each of the following shapes.
Give your answers correct to two decimal places.
  a A square of side 12.6 cm.
  b A rectangle with sides of 8.5 m and 12.2 m.
  c A trapezium of height 12 cm and parallel sides of 8.5 cm and 11.8 cm.

4 A rhombus has an area of 5600 mm² and sides of length 8 cm.
What is its perpendicular height?

5 This trapezium has been divided into four triangles by drawing in its diagonals.

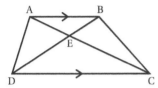

  a Show that the area of triangle ACD is equal to the area of triangle BCD.
  b Which two of the four internal triangles ABE, BEC, DEC and AED must have the same area? Explain why.

Section 2: Area of circles and sectors

HOMEWORK 16C

1 Find the area of each circle.
Use the value of π given by your calculator and give your answers correct to two decimal places.

  a 1.4 cm
  b 0.7 m
  c 2.9 m
  d 12 cm
  e 45 mm

2 Find the area of each shape. Give your answers correct to three significant figures.
  a 140 mm
  1b 14 mm
  c 20° 10 cm
  d 300° 8 cm

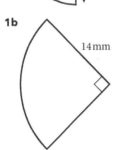
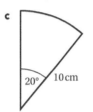
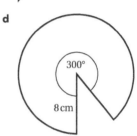

3 A mobile phone mast provides a clear signal for up to 6.5 km in all directions.
Calculate the area that has a good signal.

4 A staff room contains a rectangular table 1.3 by 0.8 m and a circular table of radius 0.55 m.
Which table top has the greatest work area?

5 A round table with a diameter of 1.2 m is covered with a circular tablecloth that extends 15 cm below the level of the table (all around it).
What is the area of the table cloth?

6 A circle has an area of 6 cm².
What is its radius?

7 A rotating water spray covers an area of 7 m². How far away from the sprayer would you need to stand to make sure you were outside the watering area?

8 A park contains two flower beds, each with an area of 31.5 m². One is circular, the other is square. Which one has the greater perimeter? Show how you worked this out.

Section 3: Area of composite shapes

HOMEWORK 16D

1 Calculate the area of each shape. (Assume that shapes **c, d, e, g** and **i** contain semicircles.)

a
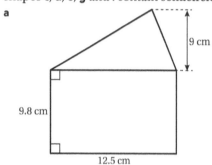

b
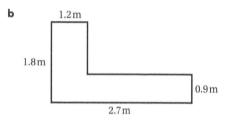

c
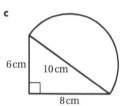

d
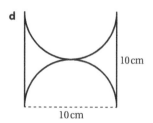

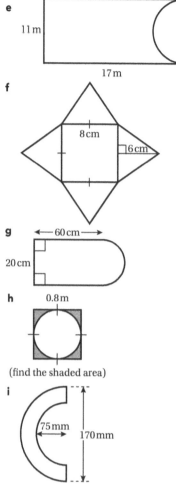

e 11 m, 17 m

f 8 cm, 6 cm

g 60 cm, 20 cm

h 0.8 m
(find the shaded area)

i 75 mm, 170 mm

2 Find the area of the shaded section of each shape. Show your working clearly. All dimensions are in centimetres.

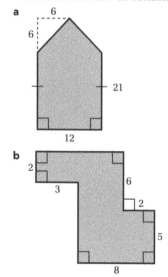

a 6, 6, 21, 12

b 2, 3, 6, 2, 5, 8

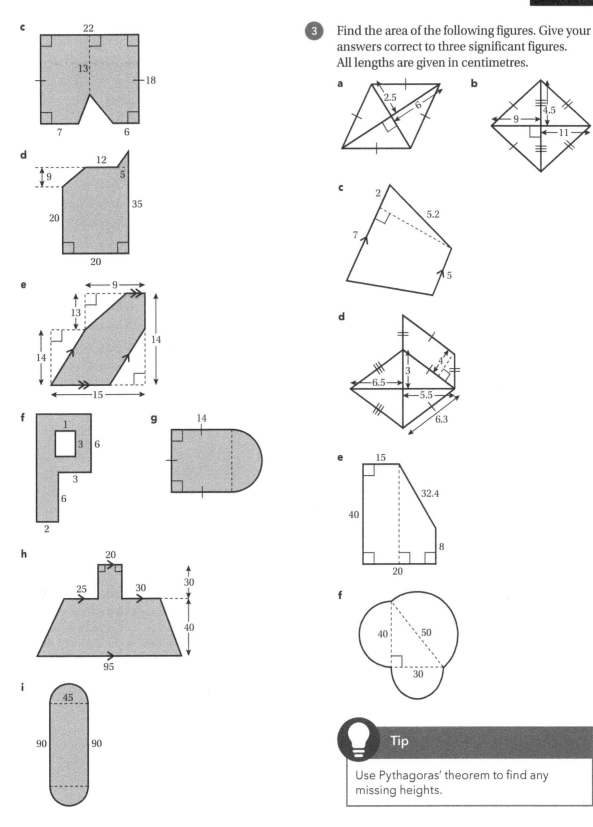

c
22
13
18
7
6

d
12
9
5
35
20
20

e
9
13
14
14
15

f
1
3
6
3
6
2

g
14

h
20
25
30
30
40
95

i
45
90
90

3. Find the area of the following figures. Give your answers correct to three significant figures. All lengths are given in centimetres.

a
2.5
6

b
4.5
9
11

c
2
5.2
7
5

d
3
4
6.5
5.5
6.3

e
15
32.4
40
8
20

f
40
50
30

Tip

Use Pythagoras' theorem to find any missing heights.

4 These shapes are all made up of circles and squares or parts of these shapes. Calculate the area of each shaded part, giving your answers correct to two decimal places. The dimensions are all in centimetres.

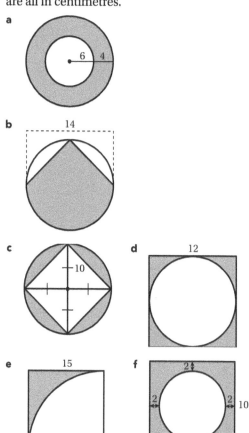

5 A 1.2 m wide path is built around a circular pond of diameter 3.8 m.
   a Determine the area of the surface of the pond.
   b Determine the area of the path.
   c The path is to be made of gravel, which is sold in bags that cover x square metres and cost £y. How much will the path cost?

6 A DVD has a diameter of 12 cm. There is a round hole in the centre of the disk and it has a diameter of 15 mm.
   a The top surface of the disk is to be printed with a logo. What is the total area available for printing?
   b The disks are packed in rectangular boxes 130 mm × 190 mm.

Determine the area of plastic visible when the disk is in place on the base of the box like this:

   c What is the total perimeter of the disk?

7 These designs are printed on square disks with an area of 30 cm². Which design requires the most grey paint?

8 A round rug of diameter 2.4 m is placed on the floor of a rectangular room 3.2 × 4.1 m. How much floor space is clear once the rug is in place?

Chapter 16 review

Use $\pi = 3.142$ for any questions involving circles.

Give all your answers to three significant figures where appropriate.

1 A circular plate on a cooker has a diameter of 21 cm. There is a metal strip around the outside of the plate.
   a Determine the area of the cooking surface of the plate.
   b What is the length of the metal strip?
   c Will a round frying pan with a base of area 397 cm² fit onto this cooking plate with no overlap? Show how you decided this.

2 What is the radius of a circle with an area of 65 cm²?

3 MNOP is a trapezium with an area of 150 m². Calculate the length of NO.

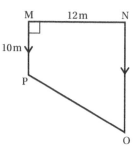

4 Calculate the shaded area of each figure.

a

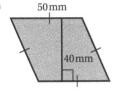

50 mm

40 mm

b

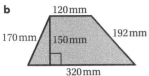

120 mm

170 mm · 150 mm · 192 mm

320 mm

c

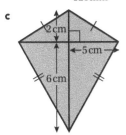

2 cm

←5 cm→

6 cm

d

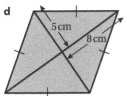

5 cm

8 cm

e

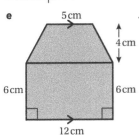

5 cm

4 cm

6 cm · 6 cm

12 cm

f

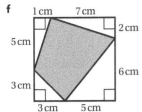

1 cm · 7 cm

2 cm

5 cm

6 cm

3 cm

3 cm · 5 cm

g

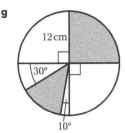

12 cm

30°

10°

5 The search area for a missing hiker has been narrowed down to the shaded area of this sector of a national park. Determine the area of the park that needs to be searched.

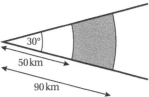

30°

50 km

90 km

6 A large circular pizza has a diameter of 25 cm. If it is cut into eight equal slices, calculate the area of each slice.

7 The small indoor pool at a community centre is rectangular with a semi-circular shallow area at one end. The rectangular part of the pool is 10 m long and 4.3 m wide. There is a 2 m wide non-slip area around the pool.

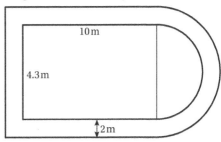

10 m

4.3 m

2 m

a What is the area of the non-slip flooring?
b Half the flooring has to be resurfaced at a cost of £132.50 per square metre. What will this cost?
c The bottom of the pool is to be painted with an antifungal paint. Determine the total floor area to be painted.
d The paint comes in 2 litre tins, each with a coverage area of 20 000 cm². How many tins will be needed to paint the floor of the pool?

17 Approximation and estimation

Section 1: Rounding

HOMEWORK 17A

1. Round these numbers to the nearest 100.
   a 2345   b 27 907   c 143

2. Round these measurements to the nearest whole unit.
   a 8.6 cm   b 13.14 m   c 0.1987 m

3. Round each value to the nearest million.
   a 13 499 000   b 1 987 654   c 21 097 099

4. Say whether each value will round to £10 or not.
   a £9.56   b £10.56   c £10.45   d £10.09

5. Estimate the length of the side of a cube with a volume of 12 cm².

HOMEWORK 17B

1. Write the following fractions as decimals correct to two decimal places.
   a $\dfrac{1}{9}$ b $\dfrac{2}{3}$ c $\dfrac{5}{9}$ d $\dfrac{3}{11}$

2. Write the following values correct to one decimal place.
   a 4.197   b 30.048   c 6.329   d 6.392

3. Round each money amount in the lists to the nearest pound.
   Write an estimated total amount for each list.

List a	List b	List c	List d
3.89	3.49	3.69	203.19
0.19	1.19	17.99	199.99
1.99	0.39	1.69	201.09
2.10	2.29	0.75	107.25
3.89	11.99	14.29	124.50
		0.99	89.04
		12.29	

4. Calculate the difference between your estimated total for each list in question 3 and the actual amount.

5. For each of the following situations, round the given values to a suitable level of accuracy.
   a Amit runs 100 m in 13 seconds. He covers about 7.692307 m in a second.
   b £7.45 is shared equally among eight people. Each person should get £0.93125.
   c A plane travels 4200 km in 7.75 hours, this is a speed of 541.9354839 km/h.
   d Apply fertiliser at a rate of 0.3947 l/m^2.

HOMEWORK 17C

1. How many significant figures are there in each of these values?
   a 345   b 12.096   c 0.00188
   d 8.0   e 0.007   f 0.120

2. Round each of these numbers to three significant figures.
   a 53 217   b 712 984
   c 17.364   d 0.007279

3. Round 24.738095 to:
   a two significant figures
   b five significant figures.

4. Round 0.0024835 to:
   a four significant figures
   b one significant figure.

5. A building supply store is selling square tiles with an area of 790 cm².
   a Is it possible to have square tiles whose area is not a square number? Explain.
   b Find the length of each side of the tile correct to three significant figures.
   c What is the minimum number of tiles you would need to tile a rectangular floor 3.6 m long and 2.4 m wide?

6. Calculate and give your answers to the number of significant figures indicated.
   a 2.1245×0.0007654 (2 sf)
   b $0.62212 \div 3.1119$ (5 sf)
   c $4.2365 \div 0.0109$ (3 sf)

HOMEWORK 17D

1 For each value below:
   a round to two decimal places.
   b truncate to two decimal places.
   c truncate to a whole number.
   d truncate after the third significant figure.

i	98.847	ii	5.4396	iii	239.364
iv	0.009786	v	2017.968	vi	4.6984
vii	0.008898	viii	125.7449	ix	5023.505
x	0.7654				

2 A computer criminal was caught stealing money from bank clients. He had adapted the programme to truncate interest amounts due on account after four decimal places. So, an interest amount of £100.089775 would transfer to the account as £100.0897 and the 0.000075 would accrue to the criminal's account. He had being doing this for 15 years before he was caught.
   a Why do you think this went unnoticed for so long?
   b How many transactions would need to be processed for the criminal to earn a pound?

Section 2: Approximation and estimation

HOMEWORK 17E

1 Use whole numbers to show why:
   a $3.9 \times 5.1 \approx 20$ **b** $68 \times 5.03 \approx 350$
   c $999 \times 6.9 \approx 7000$ **d** $42.02 \div 5.96 \approx 7$

2 Josh is paid £8.45 per hour. He normally works 38 hours per week.
   a Estimate his weekly earnings to the nearest pound.
   b Estimate approximately how much he earns in a year if he takes two weeks off for holidays.

3 Give an estimated answer for each calculation.
   a 0.82×21.75 **b** 0.816×0.207
   c 140.7×5.9 **d** 12.35×0.025
   e 12.45×8.89 **f** $(4.25 \times 12.15) \div 7.3$

4 Estimate the answers to each of these calculations to the nearest whole number.
   a 9.75×4.108 **b** $0.0387 \div 0.00732$
   c $\dfrac{39.4 \times 6.32}{9.987}$ **d** $\sqrt{64.25} \times (3.098)^2$

5 Estimate the answer to each calculation to the nearest whole number.
   a $5.2 + 16.9 - 8.9 + 7.1$
   b $(23.86 + 9.07) \div (15.99 - 4.59)$
   c $\dfrac{9.3 \times 7.6}{5.9 \times 0.95}$ **d** $(8.9)^2 \times \sqrt{8.98}$

6 Write down an approximate calculation to show that $5.78 \times £51.30$ is about £300.

7 A fast food outlet uses a ticket system to serve people. In one hour, they serve 394 customers. Each customer spends an average of £3.09. Estimate the total earnings per hour for this store.

8 **a** Find an approximate answer to $(3.802 + 7.54) \div 3.27$.
   b Calculate the exact value of $(3.802 + 7.54) \div 3.27$ correct to two significant figures.

9 Estimate by using known facts about square and cube numbers.
   a $\sqrt{11}$ **b** $\sqrt{2}$ **c** $\sqrt{8}$ **d** $\sqrt[3]{4}$ **e** $\sqrt[3]{9}$

Section 3: Limits of accuracy

HOMEWORK 17F

1 Each of the numbers below has been rounded to the degree of accuracy shown in the brackets.
   Find the upper and lower bounds in each case.
   a 42 (nearest whole number)
   b 400 (1 sf)
   c 12.24 (2 dp)
   d 2.5 (to nearest tenth)
   e 390 (nearest ten)
   f 60 cm (nearest 10 cm)
   g 5.6 cm (nearest mm)
   h 28 g (nearest gram)
   i 6.5 seconds (nearest $\frac{1}{10}$ of a second)
   j 1.23 litres (3 sf)

2 A building is 72 m tall measured to the nearest metre.
   a What are the upper and lower bounds of the building's height?
   b Is 72.499999999999999999 metres a possible height for the building? Explain why or why not.

3 Jess took 25.7 seconds to complete a maths problem. The time is correct to the nearest tenth of a second. Between which values could the actual time lie? Give your answer as an inequality using t for time.

HOMEWORK 17G

1 The dimensions of a rectangular piece of land are 4.3 m by 6.4 m. The measurements are each correct to one decimal place.
   a Find the area of the piece of land.
   b Calculate the upper and lower bounds of the area of the land.

2 A rectangle is 3.5 cm wide and 4.6 cm long measured to two significant figures. Determine the lower and upper bounds of:
   a the perimeter of the rectangle
   b the area of the rectangle.

3 Usain Bolt holds the world records for the 100 m and 200 m sprints (correct in 2014). He was also a member of the Jamaican four by 100 m relay team that set a new world record of 36.84 seconds in August 2012.
   a Usain Bolt is 196 cm tall, correct to the nearest centimetre, and his mass is 94 kg, correct to the nearest kilogram. Find the upper and lower bounds of his height and mass.
   b The Jamaican coach says his team can run the 400 m relay in 34 seconds. Both these measurements are given to two significant figures. What is the maximum speed (in metres per second) at which they can run the relay? Show how you worked out your answer. Give your answer correct to two decimal places.

4 The two short sides of a right-angled triangle are 4.7 cm (to nearest mm) and 6.5 cm (to the nearest mm). Calculate upper and lower bounds for:
   a the area of the triangle
   b the length of the hypotenuse.
   Give your answers to four decimal places.

5 Given that $y = 3\frac{x}{z}$ and $x = 3.62$ and $z = 5.41$ both correct to three significant figures, find the upper and lower bounds of y. Show how you worked out your answers. Give your answer to two decimal places.

6 The maximum height (h) of an object thrown vertically upwards can be calculated using the formula:
$$h = \frac{u^2}{2g}$$
where u is the horizontal velocity and g is gravitational pull. Given that $u = 4.2$ and $g = 9.8$, both correct to two significant figures, determine the highest and lowest possible height reached by the object in question.

7 A triangle has an area of 14.5 cm² and a perpendicular height of 4.6 cm, both given to the nearest 0.1 unit. Work out the limits between which the base length must lie. Give your answers to two decimal places.

Chapter 17 review

1 Round each of the following numbers to the accuracy shown in brackets.
   a 15.638 (1 dp)   **b** 383.452 345 (3 sf)
   c 0.000 034 556 (2 sf)   **d** 0.999 98 (3 dp)

2 Estimate the value of each of the following.
   a $\sqrt{6.1 + 2.9}$ **b** 14.6×2.7
   c $46.2 \div 25.3$ **d** $(23.4)^2$
   e 125×384 **f** $\dfrac{36.5 + 28.2}{29.9 + 4.8}$
   g $\sqrt{49.1 \times 24.8}$ **h** $\dfrac{\sqrt{99.6}}{\sqrt{143}}$

3 Tayo's height is 1.62 m, correct to the nearest cm. Calculate the least possible and greatest possible height that he could be.

4 A child is weighed at a clinic and her mass is recorded to the nearest half kilogram as 12.5 kg. What is the greatest and least possible mass of this child?

5 The dimensions of a rectangle are 3.61 cm and 2.57 cm, each correct to three significant figures.
   a Write down the range of possible values of each dimension.
   b Find the lower and upper bounds of the area of the rectangle.
   c Write down the lower and upper bounds of the area correct to three significant figures.

6. The masses of three parcels to 1 dp are 0.5 kg, 0.3 kg and 0.4 kg.
What are the upper and lower bounds of their combined mass?

7. A rectangle has an area of 90.8 m² (correct to 1 dp) and is 15.4 m long (correct to 1 dp). What are the upper and lower bounds of its width?

8. The diagram opposite shows two right-angled triangles ABC and ACD.

All measurements are given to the nearest half centimetre. Calculate upper and lower bounds for the value of x, giving your answers to 1 decimal place.

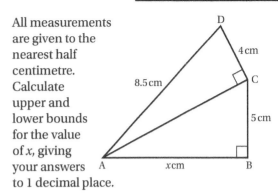

18 Straight-line graphs

Section 1: Plotting graphs
HOMEWORK 18A

1. Draw up a table of values for each equation.
Use ⁻1, 0, 1, 2 and 3 as values of x.
In question **g** use these values for y.

a $y = x + 3$ b $y = 3$
c $y = \dfrac{1}{2}x - 1$ d $y = \dfrac{-1}{2}x$
e $y = x - 1$ f $2x - y = 4$
g $x = 7$ h $x + y = ^-1$

2. Use the values from your table to plot the graphs.
Plot graphs **a** to **d** on one set of axes and graphs **e** to **h** on another.

Section 2: Using the features of straight-line graphs
HOMEWORK 18B

1. Find the gradient of each line.

a 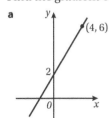 b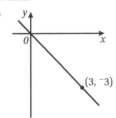

c 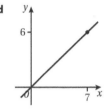 d

e 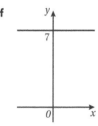 f

g 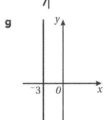 h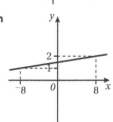

2. Determine the gradient of each of the following graphs (without drawing the graph).

a $y = x$ b $y = \dfrac{x}{2} + \dfrac{1}{4}$
c $y = 4\dfrac{x}{5} - 2$ d $y = 7$
e $y = ^-3x$ f $x + 3y = 14$
g $x + y + 4 = 0$ h $2x = 5 - y$
i $x + \dfrac{y}{2} = ^-10$

3 Determine the gradient of a line which passes through each pair of points.
  a $(0, 0)$ and $(^-3, 3)$ **b** $(4, 2)$ and $(8, 4)$
  c $(2, ^-3)$ and $(4, ^-1)$

4 **a** Draw a set of axes and plot the vertices of quadrilateral ABCD with A = $(0, 3)$, B = $(4, 5)$, C = $(2, 1)$ and D = $(^-2, ^-1)$. Draw in the sides of the quadrilateral.
  b Determine the gradient of each side of the quadrilateral.
  c What is the gradient of diagonal AC?

5 A straight line intersects the curved graph $y = x^2 - 4x$ at the points where $x = ^-1$ and $x = ^-2$.
  Determine the equation of the straight line.

HOMEWORK 18C

1 Write equations for graphs **a** to **e**.

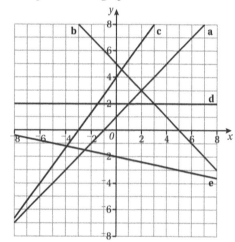

2 Use the gradient-intercept method to sketch these graphs.
  a $y = 2x + 1$ **b** $y = ^-3x + 2$
  c $y = \frac{1}{2}x + 1$ **d** $x - 4y = 2$

3 Determine the x- and y-intercept of each line and sketch the graphs.
  a $x + y = 4$ **b** $x + 2y = 6$
  c $2x - y = 4$ **d** $3x + 2y = 2$

4 Sketch the graphs of $y = 2x - 3$ and $^-3x = 6y + 6$ on the same set of axes. Mark and label the intercepts with the axes and label each graph correctly.

Section 3: Parallel lines, perpendicular lines and tangents

HOMEWORK 18D

1 Are the following pairs of graphs parallel or not?
  a $y = ^-3x$ and $y = ^-3x + 7$
  b $y = 0.8x - 7$ and $y = 8x + 2$
  c $2y = ^-3x + 2$ and $y = \frac{2}{3}x + 2$
  d $2y - 3x = 2$ and $y = ^-1.5x + 2$
  e $y = 8$ and $y = ^-9$
  f $x = ^-3$ and $x = \frac{1}{2}$

2 What is the equation of a graph parallel to $y = x + 5$ and passing through point $(0, ^-2)$?

3

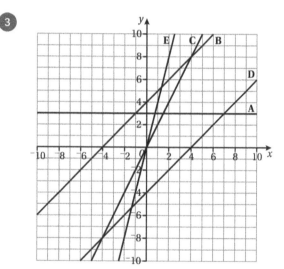

For lines A to E:
  a determine the equation of the line.
  b determine the equation of a line parallel to each line and passing through point $(0, ^-7)$.

4 If $y = ax - 8$ is parallel to $y = 7x + \frac{5}{2}$, what is the value of a?

5 If $y = bx - 1$ is parallel to $4y - 5x = 7$, what is the value of b?

6 A line passes through points A$(0, 4)$ and B$(^-3, ^-1)$. Find the equation of the line parallel to AB which passes through point $(3, 5)$.

7 Could you join points P$(^-3, 6)$, Q$(3, 2)$, R$(1, 0)$ and S$(^-4, 3)$ to draw a parallelogram PQRS? Justify your answer.

HOMEWORK 18E

1 A line perpendicular to $y = \dfrac{x}{5} + 3$ passes through $(1, 3)$.
What is the equation of the line?

2 Are these pairs of lines parallel, perpendicular or neither?
  a $2x - 4y = 6$ and $y + 2x = 4$
  b A line through $(^-2, 4)$ and $(0, 6)$ and a line through $(^-4, 3)$ and $(0, 7)$.
  c $\dfrac{(x + y)}{2} = 5$ and $x - 4 = y$

3 Two graphs are sketched on the same system of axes.

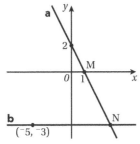

  a Determine the equations of lines a and b.
  b What are the coordinates of points M and N?
  c What is the equation of a line perpendicular to b, passing through point $(3, 9)$?

4 In the diagram, ABCD is a rectangle.

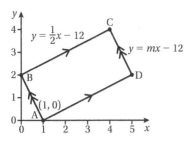

  a Determine the value of m for line CD.
  b Determine the equation of AB.
  c Show that ADC is a right angle.
  d Use the equations of lines BC and CD to determine the coordinates of point C.

5 Show that the points $A(^-3, 6)$, $B(^-12, ^-4)$ and $C(8, ^-5)$ could not be the vertices of a rectangle ABCD.

HOMEWORK 18F

1 Two tangents to circle $x^2 + y^2 = 100$ are shown on the diagram.
Determine the gradients of each tangent.

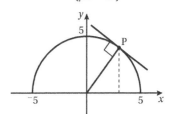

2 A tangent touches circle $x^2 + y^2 = 81$ at point $(0, 9)$. What is the equation of the tangent?

3 A line touches a circle $x^2 + y^2 = 169$ at one point $(^-12, ^-5)$ only.
What is the equation of the line?

4 The gradient of the tangent at P can be found using the formula $y = \dfrac{^-x}{(\sqrt{25 - x^2})}$.

Determine the gradient of the tangent when:
  a $x = 1$ **b** $x = 3$ **c** $x = ^-2$

Section 4: Working with straight-line graphs

HOMEWORK 18G

1 Sketch each of the graphs labelling the key features.
  a $y = 3x - 2$ **b** $y = 4x - 2$
  c $y = \dfrac{1}{4}x$ **d** $y = ^-x - \dfrac{1}{4}$
  e $y = \dfrac{3x}{4 + 1}$ **f** $2y + x = 4$

2 Find the equation of a line that is:

  a parallel to the line with equation $y = 4x + 1$, but passes through the point $(3, 16)$

  b parallel to the line with equation $y = {}^-3x + 5$, but passes through the point $(7, {}^-8)$

  c parallel to the line with equation $y = 0.5x + 0.3$, but passes through the point $(3, 2.4)$

  d parallel to the line with equation $3x + 4y = 12$, but passes through the point $(2, {}^-1)$

  e parallel to the line with equation $5x - 2y = 18$, but passes through the point $({}^-3, {}^-4)$.

  f perpendicular to the $y = {}^-\frac{1}{2}x + 2$ and passing through $(1, 1)$.

  g a tangent to circle $x^2 + y^2 = 25$, touching the circle at $(0, 5)$.

3 The point $(5, a)$ lies on the line $y = \frac{1}{2}x - 1$. What is the value of a?

4 If the point $(b, 7)$ lies on the line $y = 2x + 3$, what is b?

5 If the graph $y = 2x + 6$ passes through points $(3, m)$ and $(n, 2)$ determine the value of m and n.

6 **a** Find the equation of each line A to D in the following diagrams.

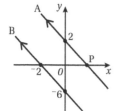 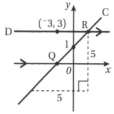

  b What are the coordinates of points P, Q and R?

  c What is the equation of the line parallel to graph **c** which passes through the origin?

7 Show that lines $y = 4x - 3$ and $x + 4y = 0$ are perpendicular.

Chapter 18 review

1 Draw up a table of values for $x = {}^-1, 0, 1, 2$ for each equation and use the tables to plot the graphs.

  a $y = \frac{1}{2}x$ **b** $y = \frac{{}^-1}{2}x + 3$

  c $y = 2$ **d** $y - 2x - 4 = 0$

2 What equation defines each of these lines?

  a A line parallel to $y = \frac{{}^-4}{5x}$ which passes through the point $(0, {}^-3)$.

  b A line parallel to $2y + 4x = 20$ with a y-intercept of ${}^-3$.

  c A line parallel to $x + y = 5$ which passes through $(1, 1)$.

  d A line parallel to the x-axis which passes through $(1, 2)$.

  e A line perpendicular to the y-axis which passes through $({}^-4, {}^-5)$.

  f A line perpendicular to $y + 4x = 2$ and passing through $(8, 3)$.

3 Given points $D(1, {}^-2)$, $E(3, 5)$, $F(8, 6)$ and $G(6, 0)$, is DEFG a parallelogram? Justify your answer.

4 Triangle ABC is drawn on a grid. AM bisects BC at M and TB is the height of the triangle from base AC.

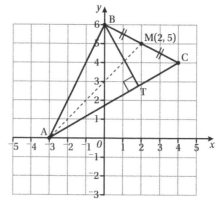

  a Determine the equation of each side of the triangle.

  b What is the equation of AM?

  c Determine the equation of BT.

  d A circle of centre $(0, 0)$ is drawn on the grid. The line from the origin to vertex C is a radius of the circle. What is the equation of the circle?

5 A rectangle and triangle are drawn on a grid.

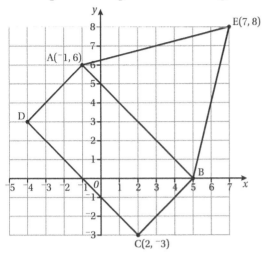

a Use gradients to show that ABC is a right angle.
b Given that AE = BE, determine the equation of the perpendicular bisector of AB, through point E.
c Show that the perpendicular bisector of AB is parallel to both BC and AD.

19 Graphs of equations and functions

Section 1: Review of linear graphs

HOMEWORK 19A

1 Write the equation of each graph, **a** to **e**.

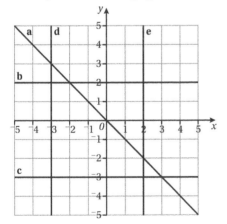

2 Write the equation of each of the following graphs.
  a A line parallel to the y-axis and passing through point $(2, 0)$.
  b The set of points with x-coordinate of $^-3$.
  c The y-axis.
  d The x-axis.
  e the line perpendicular to the x-axis at 3.
  f the line parallel to the x-axis and passing through y at $1\frac{1}{2}$.
  g the set of all points with y-coordinate of $^-1$.

3 Look at the diagram and answer the questions.

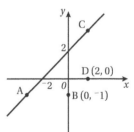

a What is the equation of line AC?
b Given that AB is parallel to the x-axis, what is the equation of a line through points A and B?
c Given that CD is perpendicular to the y-axis, what would the equation be of a line joining these points?

4 Sketch the graphs of the following linear equations marking the essential points.

a $y = 2x + 1$

b $y = \dfrac{-3x}{2} + 2$

c $x = 3$

d $3x + 2y = 6$

Section 2: Graphs of quadratic functions

HOMEWORK 19B

1 Construct a table of values, selecting values from $-6 \leqslant x \leqslant 6$, for each of the following equations. Draw the graphs on the same set of axes.

a $y = -x^2 + 3$

b $y = x^2 - 2x + 2$

c $y = x^2 + 8x + 16$

2 This is the graph of $y = x^2 - 4x + 3$.

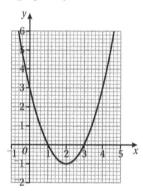

Use the graph to estimate the solution of:

a $x^2 - 4x + 3 = 0$

b $x^2 - 4x + 3 = 3$

c $x^2 - 4x = 1$

3 By drawing up a table of values, and drawing the graphs on the same set of axes, find the approximate solutions to the simultaneous equations:

$$2y = x - 1 \text{ and } y = x^2 - 4x + 3.$$

HOMEWORK 19C

1 For each graph A to E, identify:

a the turning point and whether it is maximum or minimum.

b the axis of symmetry.

c the y-intercept.　　d the x-intercepts.

A
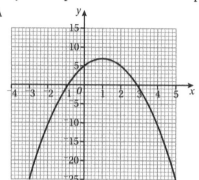

B
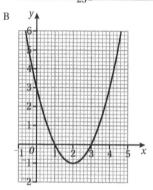

C
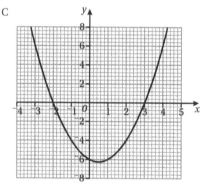

D　　　　　　　　E
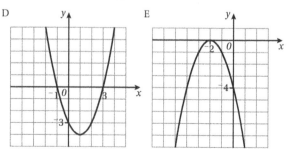

2 Each statement below this graph is **false**. Identify the mistakes and correct the statements.

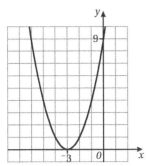

a The axis of symmetry is $y = {}^-3$.
b The turning point is a maximum at $(9, 0)$.
c The x-intercepts are $(0, 9)$ and $({}^-3, 0)$.
d The graph doesn't cut the y-axis.

3 Sketch the graph of $y = {}^-x^2 + 4x + 1$ and label its main features.

HOMEWORK 19D

1 Draw and label sketch graphs of the following.
a $y = \frac{1}{2}x^2 - \frac{1}{2}$
b $y = {}^-2x^2 + 8$
c $y = 2x^2 - 3$
d $y = \frac{1}{2}x^2 + 2$

2 Use the information on each graph to determine its equation.

a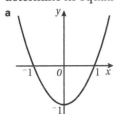
b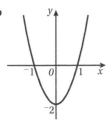
c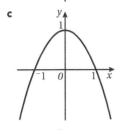
d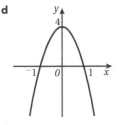
e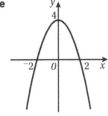
f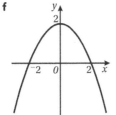

Section 3: Graphs of other polynomials and reciprocals

HOMEWORK 19E

1 Draw up a table of values for each equation. Plot each graph on a separate grid.
a $y = 3x^3$
b $y = 2x^3 - 3$
c $y = 3x^3 + 1$
d $y = \frac{1}{2}x^3 + 1$
e $y = x^3 - 4x^2$
f $y = x^3 + 2x - 10$

2 The graph of $y = 2x^3 + 2$ is shown here. Use this to draw a sketch graph showing what you would expect the graph of $y = {}^-2x^3 + 2$ to look like.

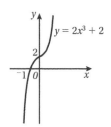

HOMEWORK 19F

1 Study the two graphs of $y = \frac{a}{x}$ below.

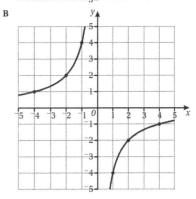

a Which graph has a positive value of a? How do you know this?
b Which graph has the line $y = {}^-x$ as its line of symmetry?
c What is the equation of each graph?

65

2 Draw up a table of values and plot each pair of graphs on the same system of axes.

a $y = \dfrac{-4}{x}$ and $y = \dfrac{-6}{x}$ 　 b $y = \dfrac{2}{x}$ and $y = \dfrac{6}{x}$

c $xy = 1$ and $xy = 4$

3 Three reciprocal graphs are shown on the grid. A point (and its reflection) is given for each graph.

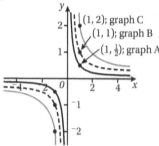

(1, 2); graph C
(1, 1); graph B
(1, ½); graph A

Match each graph to its correct equation.

$xy = 1$ 　 $2xy = 1$ 　 $xy = 2$

4 This is the graph of $y = \dfrac{4}{x}$, plotted accurately. Use the information on this graph to plot the graph of $y = \dfrac{4}{x} + 1$.

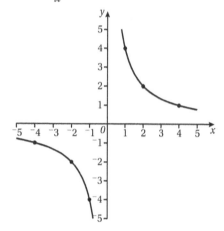

Section 4: Exponential functions

HOMEWORK 19G

1 a Draw up a table of values from ⁻3 to 3 and draw the graphs of each equation. Draw each pair on the same system of axes.

　i $y = 3^x$ and $y = \dfrac{1}{3^x}$ 　ii $y = 5^x$ and $y = \dfrac{1}{5^x}$

b Use the graphs you have drawn in part to find the solutions to the following equations:

　i $3^x = 9$ 　　ii $5^{-x} = 5$

2 Look at graph A and graph B. They are both graphs of the general form $y = bx$. What can you say about the value of b in each case?

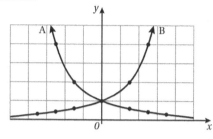

3 The graph $y = 2^{-x}$ is shown here.

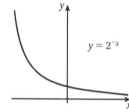

$y = 2^{-x}$

a What are the coordinates of the y-intercept?
b Give the coordinates of two other points which lie on this graph.
c Is the graph an increasing or decreasing function? Explain how you know this.
d Another graph is drawn so that the two graphs are symmetrical about the y-axis. What is the equation of the other graph?

Section 5: Circles and their equations

HOMEWORK 19H

1 Sketch and label the graphs of:

a $x^2 + y^2 = 2^2$ 　　　 b $x^2 + y^2 = 16$
c $x^2 + y^2 = 1$ 　　　 d $4x^2 + 4y^2 = 9$
e $y^2 = (1 - x)(1 + x)$ 　 f $x^2 + y^2 = a^2$

2 Does point (⁻5, 5) lie on the circle $x^2 + y^2 = 50$?

3 Write the equation of each of the following graphs.

a
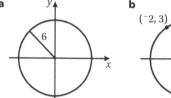

b
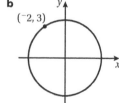
(⁻2, 3)

c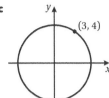

d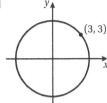

e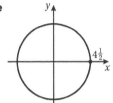

f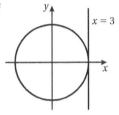

4 Three graphs are drawn on the same system of axes.

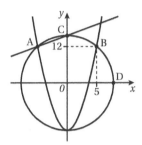

a Write an equation for:
  i the circle.
  ii the line AC.
  iii the parabola.
b Find the coordinates of A, B and C.

5 **a** Make y the subject of the equation $x^2 + y^2 = 25$.
b What do you think the graph of the rearranged equation will look like? Why?
c Sketch the graph of $y = ^-\sqrt{(1 - x^2)}$.

Chapter 19 review

1 Read each statement. Decide whether it is true or false. If it is false, write a correct version.
  a The graph of $xy = k$ is the same as the graph of $y = \dfrac{k}{x}$.
  b $2x^3 - 3y + 1 = 0$ is a straight line graph with a gradient of $\dfrac{2}{3}$.
  c The graph $x = k$ is a straight line parallel to the x-axis.
  d The standard equation $y = ax^3 + bx + c$ will produce a cubic curve.
  e $y = 3x^2$ is a U-shaped graph with the y-axis as its axis of symmetry.
  f The graph of $y = \dfrac{a}{x}$ will produce an exponential curve.
  g The graph of $x^2 + y^2 = 3$ will produce a circle of radius 3.

2 Sketch each of the following graphs.
  a $y = ^-x + 2$ **b** $y - \dfrac{^-4}{x}$
  c $y = 3^x$ **d** $y = ^-x^2 + 3$
  e $y = \tan x$ (for values from 0 to 360)

3 Determine the equation used to generate each of the following graphs.

a

b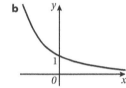

c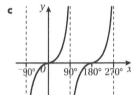

d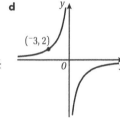

e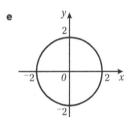

f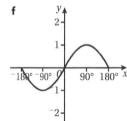

g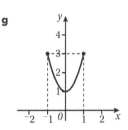

67

4 Determine the equation of the following graphs:
  a the reciprocal curve passing through (2, 3).
  b the linear graph passing through point (2, ⁻4) which is perpendicular to $y = 4 - x$.
  c a graph in the form of $y = ax^2 + q$ passing through points (2, 0) and 1,6).
  d the equation of a circle (Graph A) centred on (0,0) and a reciprocal curve (Graph B) which intersects with the circle at point (2, 4).

5 If A is (1, ⁻4) and B is (⁻5, 2), determine:
  a the equation of the line passing through the two points A and B.
  b the equation of a line parallel to AB passing through (0, ⁻2).
  c the reciprocal curve on which point B lies.
  d the straight line parallel to the y-axis which cuts the curve at B.

20 Three-dimensional shapes

Section 1: Review of 3D solids

HOMEWORK 20A

1 What shape/s are the faces of each of the following solids?
  a a cube
  b a pentagonal-based pyramid
  c a pentagonal prism
  d a cylinder
  e a trapezoidal prism

2 What are the names of the following shapes?

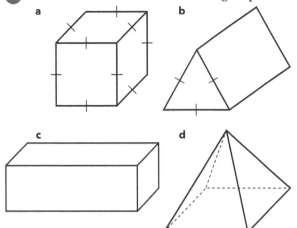

3 Name the shape given the following properties.
  a A four-sided shape with two pairs of equal and opposite sides but no right angles.
  b A four-sided shape with only one pair of parallel sides.
  c A triangle with two equal angles.
  d A triangle with all sides and angles equal.
  e A four-sided shape with two pairs of equal and adjacent sides.

Section 2: Drawing 3D objects

HOMEWORK 20B

1 Draw the following solids on centimetre squared grid paper.
The dimensions are given in centimetres.

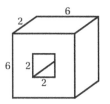
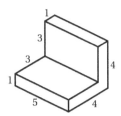

2 Draw these shapes on isometric grid paper.
The dimensions are given as distances between the dots on the paper.

3 Assuming no blocks are missing, how many blocks would you need to build each of the following solids?

a   b

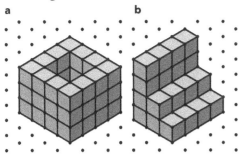

4 Copy and complete diagram B to show the same object as diagram A from the given orientation.

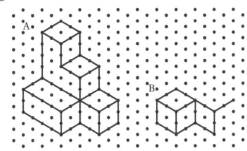

5 In each of the following solids, block X is moved and placed on top of block Y. Draw the resultant solids on isometric grid paper.

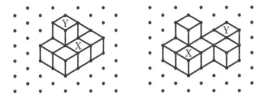

Section 3: Plan and elevation views
HOMEWORK 20C

1 Draw a plan view, front elevation and right-hand side view of each of the following solids.

2 Draw the plan, front and right-hand view of this building.

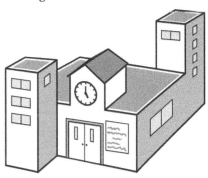

Chapter 20 review

1 Sketch and label the net of:
   a a cube with edges 4 cm long
   b a cuboid with a square face of side 2 cm and a length of 4 cm.

2 This solid has been drawn on a squared grid. Redraw it on an isometric grid.

3 **a** Draw this solid on an isometric grid.
   b Draw the plan view, front elevation and right-hand side elevation of this solid.

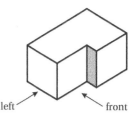

left   front

4 Draw an accurate plan view and left, front and right elevation of this solid built from 1 cm² cubes.

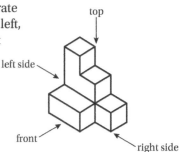

top

left side

front

right side

21 Volume and surface area

Section 1: Prisms and cylinders
HOMEWORK 21A

Give your answers to three significant figures where appropriate.

1 Calculate the volume of each prism.

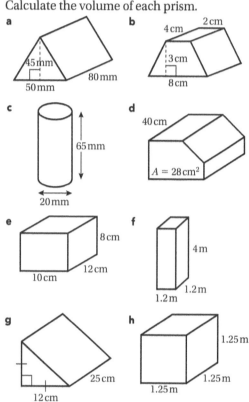

2 A pocket dictionary is 14 cm long, 9.5 cm wide and 2.5 cm thick.
Calculate the volume of space it takes up.

Tip

It is useful to draw a rough net of the object to make sure you include all the faces in your surface area calculations.

3 A wooden cube has six identical faces each of area 64 cm².
  a What is the total surface area of the cube?
  b What is the height of the cube?

4 A teacher is ordering wooden blocks to use in her maths classroom. The blocks are cuboids with dimensions 10 cm × 8 cm × 5 cm.
  a Calculate the surface area of one block.
  b The teacher needs 450 blocks. What is the total surface area of all the blocks?
  c The blocks are to be varnished. A tin of varnish covers an area of 4 m². How many tins of varnish are needed to coat each block once? Show how you worked out your answer.

5 The figure shows a metal canister with a plastic lid. Calculate:
  a the volume of the canister.
  b the surface area of the outside of the metal canister.
  c the area of the top of the plastic lid.

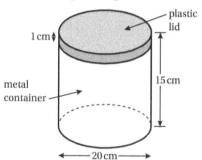

6 The radius of a cylinder is 90 cm. Its height is double its radius. What is its surface area? Give your answer correct to three significant figures.

7 A rectangular box measures 280 mm × 140 mm × 150 mm.
What is the maximum number of smaller cuboids measuring 10 mm × 10 mm × 20 mm that could be packed into the box?

Section 2: Cones and spheres
HOMEWORK 21B

1. Calculate the volume and total surface area of each of the following solids.
Give your answers to three significant figures.

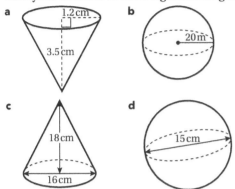

2. Find the height of a cone with a volume of 2500 cm³ and a base of radius 10 cm.
Give your answer to three significant figures.

3. Determine the volume of half a sphere of radius 6 cm.
Give your answer to three significant figures.

4. A spherical ball has a surface area of 500 cm³. What is the diameter of the ball?

5. Sphere A has radius r cm and sphere B has radius kr cm, where k is any positive number.

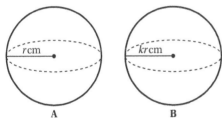

Show that:
a the surface area of sphere B is k^2 times the surface area of sphere A.
b the volume of sphere B is k^3 times the volume of sphere A.
You are now told that the volume of sphere B is 64 times the volume of sphere A.
c What is the value of surface area of A/surface area of B?
d If the volume of sphere B is p times the volume of sphere A, calculate the value of surface area of A/surface area of B.

6. What is the smallest possible diameter that a spherical metal ball could have if it has a volume greater than 1000 cm³?

HOMEWORK 21C

1. A metal ball is placed in a cylinder of water. The height of the water rises to 30 cm once the ball is placed in the cylinder. Determine the volume of the water.

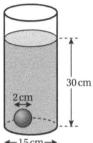

Give your answer to three significant figures.

2. A space exploration firm made this rocket by combining a cone and a cylinder.

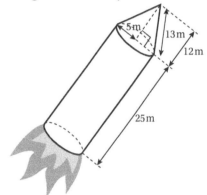

a Determine the exterior surface area of the rocket. (Do not include the base of the rocket.)
b Find the volume of the rocket.
Give your answers to three significant figures.

3. Calculate the volume of metal in this component.

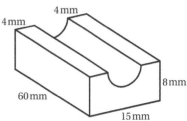

4 This is a diagram of a metal breadbin. Determine:
a the volume of the breadbin.
b its total external area.

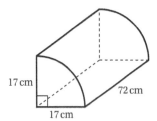

17 cm
72 cm
17 cm

5 Li has a wax cone 20 cm wide at its base and 35 cm high.
He carves a hemisphere of diameter 10 cm out of the base of the cone.
What volume of wax remains?

Section 3: Pyramids
HOMEWORK 21D

1 Determine the volume of each pyramid.

a

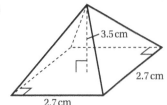

3.5 cm
2.7 cm
2.7 cm

b

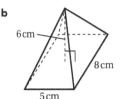

6 cm
8 cm
5 cm

c

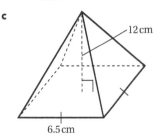

12 cm
6.5 cm

2 A rectangular building has a pyramid-shaped roof. The dimensions of the building are given on the diagram.
Calculate the volume of air inside the building.

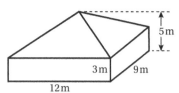

5 m
3 m
9 m
12 m

3 A pyramid has a rectangular base of area 27.9 mm². The apex of the pyramid is 9.3 mm above its base. What is its volume?

4 A triangular-based pyramid is 10.5 m tall. The base is a right-angled triangle with sides adjacent to the right angle measuring 26.6 m and 16.8 m.
What is the volume of the pyramid?

5 Sally wants to make a metal pyramid of volume 500 cm³. She starts with a square base 12 cm × 12 cm.
How high will her pyramid be?

6 Two hexagonal-based pyramids are glued together base to base.
If the area of the base is 3.6 cm² and the length from apex to apex of the two pyramids is 4.2 cm, what is the volume of the double pyramid shape?

7 A 3 m high square-based pyramid is placed on top of a 6 m high cube of concrete. The base of the pyramid fits the face of the cube exactly.
Determine the volume of the structure.

8 The Great Pyramid at Giza in Egypt has a square base of side 230 m. Its perpendicular height is 146 m.
a Determine the volume of the pyramid. Give your answer to three significant figures.
b The pyramid used to be covered in smooth stone. Given a slant height of 186.4 m, what area was covered by smooth stone?

Chapter 21 review

1 Calculate the volume and surface area of each solid.

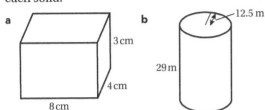

a
3 cm
4 cm
8 cm

b
12.5 m
29 m

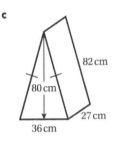

c
82 cm
80 cm
27 cm
36 cm

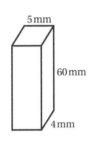

d
5 mm
60 mm
4 mm

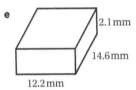

e
2.1 mm
14.6 mm
12.2 mm

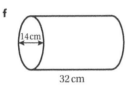

f
14 cm
32 cm

2 Here are two prisms.

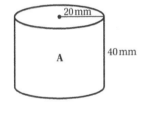

20 mm
A
40 mm

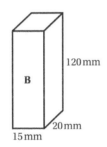

B
120 mm
15 mm
20 mm

a Which of the two prisms has the smaller volume? Show how you worked out your answer.

b What is the difference in volume?

c Sketch a net of the cuboid. Your net does not need to be to scale, but you must indicate the dimensions of each face on the net.

d Calculate the surface area of each prism.

3 How many cubes of side 4 cm can be packed into a wooden box measuring 32 cm × 16 cm × 8 cm?

4 **a** Find the volume of a lecture room that is 8 m long, 8 m wide and 3.5 m high.

b Safety regulations state that during an hour-long lecture each person in the room must have 5 m³ of air. Calculate the maximum number of people who can attend an hour-long lecture.

5 A cylindrical tank is 30 m high with an inner radius of 150 cm. Calculate how much water the tank will hold when full. Give your answer to the nearest whole number in:

a m³.

b litres.

6 A machine shop has four different cuboids of volume 64 000 mm³.
Copy and fill in the possible dimensions for each prism to complete the table.

Volume (mm³)	64 000	64 000	64 000	64 000
Length (mm)	80	50		
Width (mm)	40		80	
Height (mm)				16

7 Find the difference between the volume of a 10 cm high cone which has a 3 cm wide base and a square-based pyramid that is 3 cm wide at its base and 10 cm high.

8 Tennis balls of diameter 9 cm are packed into cylindrical metal tubes that are sealed at both ends. The inside of the tube has an internal diameter of 9.2 cm and the tube is 28 cm long.
Calculate the volume of space left in the tube if three tennis balls are packed into it.

9 For what value of r is the volume of a sphere equal to its surface area?

22 Calculations with ratio

Section 1: Introducing ratios
HOMEWORK 22A

1 Express the following as ratios in their simplest form.
   a $120:150$
   b $2\frac{3}{4}:3\frac{2}{3}$
   c 600 g to three kilograms
   d 50 mm to a metre
   e 12.5 g to 50 g
   f 3 cm to 25 mm
   g 200 ml to 3 l

2 Find the value of x in each of the following.
   a $2:3 = 6:x$
   b $2:5 = x:10$
   c $10:15 = x:6$
   d $\frac{2}{7} = \frac{x}{4}$
   e $\frac{5}{x} = \frac{16}{6}$
   f $\frac{x}{4} = \frac{10}{15}$

3 Write a ratio to compare the salaries of Nisha, Pete and Lara if Nisha earns £40 000 per year, Pete earns £35 000 per year and Lara earns £60 000 per year.

4 A triangle has sides XY = 1.2 cm, XZ = 1.6 cm and YZ = 2.0 cm.
   Determine the ratio of the sides XY : XZ : YZ in its simplest form.

5 Diego and Raheem are in the same basketball team. In one season Diego scored six more points than Raheem. Write the ratio of the number of points scored by Diego to the number of points scored by Raheem if:
   a Raheem scored 42 points.
   b Diego scored 18 points.

6 $\frac{3}{5}$ of the students in a class take French and $\frac{1}{4}$ take Spanish. Find the ratio of those who take French to those who take Spanish.

Tip

Think about how you can use ratios to compare the amount spent to the profit.

7 Phone-me-please spends £15 000 on advertising and makes a profit of £120 000. Call-me-quick spends £25 000 on advertising and makes a profit of £200 000. Which company gets the best return on their advertising spend?

8 The distance between two points on a map with a scale of 1 : 2 000 000 is 120 mm.
   What is the distance between the two points in reality? Give your answer in kilometres.

9 A plan is drawn using a scale of 1 : 500. If the length of a wall on the plan is 6 cm, how long is the wall in reality?

10 Mary has a rectangular picture 35 mm wide and 37 mm high. She enlarges it on the photocopier so that the enlargement is 14 cm wide.
   a What is the scale factor of the enlargement?
   b What is the height of her enlarged picture?
   c In the original picture, a fence was 30 mm long. How long will this fence be on the enlarged picture?

Section 2: Sharing in a given ratio
HOMEWORK 22B

1 A length of rope 160 cm long must be cut into two parts so that the lengths are in the ratio 3 : 5. What are the lengths of the parts?

2 To make salad dressing you mix oil and vinegar in the ratio 2 : 3. Calculate how much oil and how much vinegar you will need to make the following amounts of salad dressing:
   a 50 ml.
   b 600 ml.
   c 750 ml.

3 Concrete is made by mixing stone, sand and cement in the ratio 3 : 2 : 1.
   a If 14 wheelbarrows of sand are used, how much stone and cement is needed?
   b How much sand would you need if you were using 18.5 bags of stone?
   c If the concrete contains 3.8 kg of sand, what is the total mass of stone, sand and cement in the mix?

4 The size of three angles of a triangle are in the ratio $A:B:C = 2:1:3$.
What is the size of each angle?

5 A metal disk consists of three parts silver and two parts copper.
  a If the disk has a mass of 1350 mg, how much silver does it contain?
  b If a disk is found to contain 0.8 grams of silver, how much copper does it contain?

6 In a bag of berry flavoured sweets the ratio of black sweets to red sweets is $3:4$.
If there are 147 sweets in the bag, how many of them are black?

7 Statistically, the ratio $1:2$ represents people who follow a healthy lifestyle to those who are not concerned about health and whose lifestyle may not promote good health.
  a In a group of 1000 people, approximately how many people will follow a healthy lifestyle?
  b In the same group of 1000 people, about 50% of those who do not follow a healthy lifestyle eat junk food on a regular basis. Write a ratio to represent the number of people who regularly eat junk food to those who don't.

8 A company and its employees contribute to a retirement benefit fund in the ratio $11:14$.
  a If Sheldon's monthly contribution is £66, what is the company's contribution?
  b If £900 was contributed in total for Penny for a year, how much did she contribute and how much did the company contribute per month?

Section 3: Comparing ratios
HOMEWORK 22C

1 Write these ratios in their simplest form.
  a $4:9$ **b** $400\,\text{m}:1.3\,\text{km}$
  c $50\text{ minutes}:1\frac{1}{2}\text{ hours}$

2 Write these ratios in the form of $n:1$.
  a $12:8$ **b** $2\,\text{m}:40\,\text{cm}$
  c $2.5\,\text{g}$ to $500\,\text{mg}$

3 The ratio of cups of flour to number of cupcakes for two different recipes is A $1:13$ and B $2\frac{1}{2}:32$.
  a Which recipe uses the least flour per cupcake?

  b For the ratio B, determine how many complete cupcakes can be made with one cup of flour.

4 Amira mixes a drink in the ratio concentrate : water $= 1:3$. Jayne mixes her drink in the ratio $7:20$. Which mixture gives a stronger concentration?

5 Petar's mark in a test is $\frac{53}{80}$. What will his mark be if it is changed to an equivalent mark out of 50?

6 The ratio of miles to kilometres is $1:1.6093$.
  a Draw a graph to show this relationship.
  b What is the ratio of kilometres to miles in the form of $1:n$?

7 A dessert is made by mixing cream and ice-cream in the ratio $5:2$. A finished dessert contains 400 ml of cream, how much ice-cream does it contain?

8 A microscopic organism is drawn using a scale of $1:0.01$. If the organism is 60 mm long on the diagram, what is its real length?

9 Sarah worked three days a week and earned £600 per month. If she changed to working five days per week, what should her new earnings be?

10 Miguel makes a scale drawing to solve a trigonometry problem. 1 cm on his drawing represents 2 m in real life. He wants to show a 10 m long ladder placed 7 m from the foot of a wall.
  a What length will the ladder be in the diagram?
  b How far will it be from the foot of the wall in the diagram?

11 The surface area and the volume of a sphere are in the ratio $1:1$.
Calculate the radius of the sphere.

Chapter 22 review

1 Express the following as ratios in simplest form.
  a $3\frac{1}{2}:4\frac{3}{4}$
  b $5\,\text{ml}$ to 2.5 litres
  c $125\,\text{g}$ to 1 kilogram

2 Divide 600 in the ratio:
  a $7:3$ **b** $7:5$ **c** $7:13$ **d** $7:7$

3 A triangle of perimeter 360 mm has side lengths in the ratio $3:5:4$.
  a Find the lengths of the sides.
  b Is the triangle right-angled? Give a reason for your answer.

4 A model of a car is built to a scale of $1:50$. If the real car is 2.5 m long, what is the length of the model in centimetres?

5 A school computer lab is one and half times larger than a normal classroom. The locked store room in the computer lab is $\frac{1}{3}$ of the size of a normal classroom. Write a ratio in simplest form to compare the size of the computer lab to the size of the store room.

23 Basic probability and experiments

Section 1: Review of probability concepts

HOMEWORK 23A

1 In an experiment, one student rolled a dice 36 times and another rolled the dice 480 times. The outcomes of both experiments were summarised in a table.

Experiment A

Possible Outcomes	1	2	3	4	5	6
Frequency	7	6	4	6	5	8

Experiment B

Possible Outcomes	1	2	3	4	5	6
Frequency	80	78	84	78	76	84

  a Does experiment A suggest that you will get a six twice as often as a three when you roll the dice? Explain your answer.
  b If Student A rolled the dice another 36 times, how many more sixes would you expect her to get? Why?
  c The probability of rolling any number on a fair unbiased dice is $\frac{1}{6}$. Do the results of the experiment show this? Explain.
  d If you repeated experiment B, what results would you expect? Why?
  e If the student who did experiment B rolled the dice another 480 times would you expect the frequencies of each number to double? Explain why or why not.

2 An A & E at a large hospital treats 1200 patients per week. A sample of 50 patients showed that 27 were male.
  a What percentage of the sample was female?
  b How many of the 1200 patients would you expect to be male? Why?

3 Nadia made a spinner with green, red and black sectors. When she spun it 200 times she found it landed on green 60 times, on red 80 times and on black 60 times.
  a Draw a diagram to show what the spinner is likely to look like.
  b What is the probability that the spinner will land on red?

4 Research has shown the probability of a person being left-handed is 0.23. How many left-handed people would you expect to find in a population of 20 000?

5 In a sample of 100 drivers passing through a village, 17 were found to be speeding. Express this as a probability.

6 Salma has a bag containing one red, one white and one green ball. She draws a ball at random and replaces it before drawing again. She repeats this 50 times. She uses a tally table to record the outcomes of her experiment.

Red	ТНL ТНL ТНL
White	ТНL ТНL ТНL III
Green	ТНL ТНL ТНL II

a Calculate the relative frequency of drawing each colour.
b Express her chance of drawing a red ball as a percentage.
c What is the sum of the three relative frequencies?
d What should your chances be in theory of drawing each colour?

HOMEWORK 23B

1 An unbiased six-sided dice with the numbers one to six on the faces is rolled.
a What are the possible outcomes of this event?
b Calculate the probability of rolling a prime number.
c What is the probability of rolling an even number?
d What is the probability of rolling a number greater than seven?

2 Sally has ten identical cards numbered one to ten. She draws a card at random and records the number on it.
a What are the possible outcomes for this event?
b Calculate the probability that Sally will draw:
  i the number five.
  ii any one of the ten numbers.
  iii a multiple of three.
  iv a number < 4.
  v a number < 5.
  vi a number < 6.

3 There are five cups of coffee on a tray. Two of them contain sugar.
a What are your chances of choosing a cup with sugar in it?
b Which choice would you expect? Why?

4 A dartboard is divided into 20 sectors numbered from 1 to 20. If a dart is equally likely to land in any of these sectors, calculate:
a P(<8).
b P(odd).
c P(prime).
d P(multiple of 3).
e P(multiple of 5).

5 A school has 40 classrooms numbered from 1 to 40. Work out the probability that a classroom number has the numeral 1 in it.

Section 2: Further probability
HOMEWORK 23C

1 The probability that a driver is speeding on a stretch of road is 0.27. What is the probability that a driver is not speeding?

2 For a fly-fishing competition, the organisers place 45 trout, 30 salmon and 15 pike in a small dam.
a What is an angler's chance of catching a salmon on her first attempt? (Each type of fish is equally likely to respond to a fly.)
b If an angler catches a pike on her first attempt and the pike is not replaced, what is the probability of her catching another pike on her second attempt?
c Joe gets a turn to fish after two trout, four salmon and a pike have been caught and not replaced. Does he have more than a 50% chance of catching a trout? Explain how you worked out your answer.

3 Here is a list of six possible outcomes when a person is chosen at random from a large group.
Outcome A: the person is female.
Outcome B: the person is male.
Outcome C: the person is under 18.
Outcome D: the person is over 21.
Outcome E: the person has a driver's licence.
Outcome F: the person is multilingual.
Say whether each of the following pairs of outcomes are mutually exclusive or not.
a Outcomes A and B. b Outcomes A and C.
c Outcomes C and D. d Outcomes D and F.
e Outcomes E and D. f Outcomes A and E.
g Outcomes C and F.

4 In question 3 what is the maximum amount of outcomes that could be met when one person is drawn at random from the crowd? Explain your answer.

5 Andy has 1200 songs on his music player – 480 are heavy metal, 240 are drum and bass and the rest are pop.
a If he puts the player on random play, what are the chances that the first song played will be a pop song?
b What is the probability that a random song will not be heavy metal?

HOMEWORK 23D

1 The number of students who do and do not wear glasses or contact lenses is recorded in the table.

Gender	Wear glasses or contact lenses	Don't wear glasses or contact lenses
Female	116	464
Male	92	328

a Draw a frequency tree to show this data.
b What percentage of female students wear glasses?
c In a group of 100 mixed male and female students, how many would you expect to be wearing glasses?

2 In January, the weather service forecast snow on ten days of the month and no snow on the other days. It did not snow on one of the days when snow was forecast and it snowed twice on days when no snow was forecast.
a Draw a frequency tree to show this information.
b Milla says the weather forecast was accurate 90% of the time. Is she correct? Justify your answer.

3 In a survey of 250 teenagers who use online social media sites, 195 students said they were sure their passwords were secure, the others answered that they were not sure. Of those who felt their passwords were secure, 92 passwords were considered non-secure. Of those who were not sure, 32 had secure passwords.
a Show the results of this survey on a frequency tree.
b What percentage of students had secure passwords?
c What percentage of students who thought their passwords were secure actually had non-secure passwords?

4 In a global survey of 4400 parents of 14–17 year olds 44% of parents admitted that they spied on their children's social media accounts. Of this sample, 10% of the parents were British and 294 of them said they spied on their children's accounts. Of the rest of the parents, 2218 said they did not spy on their children's accounts.

a Draw a frequency tree to show the information.
b Comment on the data for British parents compared to the sample as a whole.

Section 3: Working with probability
HOMEWORK 23E

1 In an opinion poll, 5000 teenagers were asked what make of mobile phone they would choose from four options (A, B, C or D). The probability of choosing each option is given in the table.

Phone	A	B	C	D
P(OPTION)	0.36	0.12	0.4	

a Calculate P(D).
b What is P(not D)?
c What is the probability a teenager would choose either B or D?
d How many teenagers in a group of 1000 would you expect to choose Option C if these probabilities are correct?

2 Mina's dad is told by his doctor that his risk of getting heart disease in the next five years is 14.6%.
a Do you think her dad is likely or unlikely to get heart disease? Why?
b In a group of 250 people with the same risk as Mina's dad, how many would you expect to get heart disease in the next five years?

3 In a car park there are 35 red, 42 white, 12 black and 29 silver cars. 24 parking spaces are empty. What is the probability that a parking space chosen at random will contain:
a a red car?   b a silver car?
c not a black car?   d no car at all?

4 Draw unbiased spinners that will land on blue, given the following information:
a $P(blue) = \frac{1}{6}$, $P(red) = \frac{5}{6}$
b $P(blue) = \frac{1}{3}$, $P(white) = \frac{1}{3}$, $P(black) = \frac{1}{3}$
c $P(not\ blue) = \frac{1}{8}$
d $P(black) = \frac{4}{5}$, $P(blue) = P(not\ black)$.

5 A company has 1800 employees. Of those, 6% use illegal substances. The company tests all employees for illegal substances and finds that 1% of the users test negative and 1% of the non-users test positive.

   a Draw up a two-way table to show this information using actual numbers of employees.

   b Draw a frequency tree to show the data.

Chapter 23 review

1 Mia has a spinner divided into four equal sectors coloured red, yellow, green and blue.

   a If Mia spins the spinner 80 times, how many times would you expect it to land on blue? Why?

   b Mia finds that the spinner lands on blue 17 times in the 80 trials. Comment on what this result shows.

2 The table shows the actual frequency of each number on a dice in an experiment.

Outcome	Frequency in 120 trials	Relative frequency
1	16	
2	22	
3	16	
4	24	
5	24	
6	18	
Odd		
Even		
> 3		
< 5		

   a Use the information in the table to calculate the relative frequency of getting each number from 1 to 6. Give your answer as a decimal correct to two decimal places.

   b Work out the actual frequency of odd, even, > 3 and < 5 from the data given.

   c Which of these four events has the highest relative frequency?

   d If you rolled the same dice another 60 times, how many times would you expect to get a result < 5? Show how you worked out your answer.

3 The frequency tree below shows the results of a test to see whether people are allergic to cat hair.

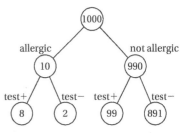

   a What percentage of people are allergic to cat hair?

   b What percentage of people test positive for the allergy?

   c What percentage of people with a positive test result are actually allergic to cat hair?

   d If a group of 50 people tested positive for cat hair allergy, how many would you expect to be allergic to cat hair?

   e What is the likelihood that you will test negative for cat allergy but actually be allergic to cats?

4 An educational authority produces the following frequency tree based on research done with GCSE students.

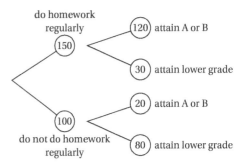

   a How many students are in the sample?

   b What percentage of students do homework regularly?

   c What percentage of students who don't do homework regularly manage to get an A or B grade?

   d In a class of 40 students, how many would you expect to attain a lower grade if they fall into the group that:

     i does homework regularly.

     ii doesn't do homework regularly.

   e What does this frequency diagram suggest?

24 Combined events and probability diagrams

Section 1: Representing combined events

HOMEWORK 24A

1 Jess has three green cards numbered one to three and three yellow cards also numbered one to three.
   a Draw up a table to show all possible outcomes when one green and one yellow card is chosen at random.
   b How many possible outcomes are there?
   c What is the probability that the number on the cards will be the same? Give your answer as a fraction in its simplest form.
   d What is the probability of getting a total < 4 if the scores on the cards are added?

2 Nick and Bev use these two spinners to work out how many moves they can make in a game. They add the scores on the spinners to get the number of moves.

 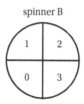

spinner A   spinner B

   a Draw up a two-way table to show all the possible pairs of numbers they can get.
   b Are all these pairs equally likely? Give a reason for your answer.
   c Calculate all the possible numbers of moves they can get by adding the scores.
   d Are all numbers of moves equally likely? Give a reason for your answer.
   e What number of moves is most likely?
   f What is the probability of this number of moves coming up at any turn?
   g What is the probability that a player will move:
      i six places.   **ii** three places.

3 A new car comes in three colours: red, white and black. The upholstery can be leather or fabric and there is a choice between a two-door or four-door model. Use a tree diagram to show all the possible options for choosing a new car.

4 Maire has a bag with a blue, red and two green counters in it. She also has a card with X on one side and Y on the other. She draws a counter at random and flips the card to land on a letter. Draw a tree diagram to show all possible outcomes.
   a What is the probability of a green counter and the letter X?
   b What is the probability of a red counter and the letter Y?

5 A tin contains eight green and four yellow counters. One counter is drawn at random, replaced and then another is drawn.
   a Draw a tree diagram to show the possible outcomes for drawing two counters.
   b Label the branches to show the probability of each event.

HOMEWORK 24B

1 In a group of 12 teenagers, 7 had a smartphone, and 8 had a tablet. 2 students had neither a smartphone nor a tablet.
Draw a Venn diagram to represent the data and state how many students had both a smartphone and a tablet.

2 A group of 24 tourists to London were asked which three attractions they had been to. The attractions were the Science Museum, the London Eye and Madame Tussaud's.
1 person had been to all three places.
3 people had visited the London Eye and Madame Tussaud's but not the Science Museum.
2 people had visited the Science Museum and Madame Tussaud's but not the London Eye.

4 people in total had been to the London Eye.
16 people in total had been to the Science
Museum.
3 of the tourists had been to none of these places.

a Copy and complete the following Venn
diagram to show the information above.

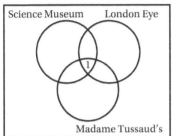

b How many of the tourists went to the
Science Museum only?

c How many people only went to the London
Eye?

d How many people only visited Madame
Tussaud's?

e Is it correct to say that one in eight tourists
visited none of these places?
Explain your answer.

Section 2: Theoretical probability of combined events

HOMEWORK 24C

Tip

For these questions, try drawing sample
space diagrams to show all possible
combinations of options.

1 Naresh has five pairs of black jeans, four black
T-shirts and three pairs of black shoes. (He likes
black.)
How many different ways are there for him to
combine these to wear a pair of jeans, a T-shirt
and a pair of shoes?

2 In how many different ways can you combine
the letters MNOP to make groups of four letters
if letters can be repeated?

3 A basketball team plays six matches in a
competition. They can win, draw or lose.
How many possible results are there?

4 A PIN number on a debit card must consist of
five digits (0–9).

a How many different PIN numbers are
possible if no digits can be repeated?

b Explain how you would work out the
number of different possible PIN numbers if
digits can be repeated.

5 How many possible outcomes are there when
you toss a coin six times in a row?

6 A multiple choice test has 20 questions.
Each question has four possible answers, A, B,
C or D. How many possible answer variations
are there?

HOMEWORK 24D

1 A blue six-sided die and a red six-sided die are
tossed together and the scores on the dice are
added to get a total.

a Draw up a grid to show all possible scores.
Determine the probability of:

b a total of 12.

c a total of 9.

d a total of at least 10.

e the score is formed by a double.

f you get a double and a score of at least 8.

2 A bag contains four green, two black and a
yellow counter. A counter is drawn, replaced
and then another is drawn. What is the
probability that:

a both counters are yellow.

b both counters are green.

c the first counter is green and the second
is black

d the counters are yellow and green in
any order?

3 Josh has a six-sided dice with the faces
painted so that three are white, two are red
and one is black. He rolls the dice and flips an
unbiased coin.

a Draw a probability space diagram to show
the possible outcomes.

b Determine the following:

  i P(red, head).

  ii P(white, tail).

  iii P(black, head).

4 Maria has a bag containing 18 fruit drop sweets. 10 are apple flavoured and 8 are blackberry flavoured. She chooses a sweet at random and eats it. Then she chooses another sweet at random. Calculate the probability that:
   a both sweets were apple flavoured.
   b both sweets were blackberry flavoured.
   c the first was apple and the second was blackberry.
   d the first was blackberry and the second was apple.
   e Your answers to a, b, c and d should add up to 1. Explain why this is the case.

5 A coloured ball is drawn from a bag and then a coin is tossed once or twice, depending on the colour of the ball drawn. There are three blue balls, two yellow balls and a black ball in the bag. The tree diagram shows the possible outcomes.

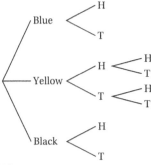

   a Copy and label the diagram to show the probability of each event. Assume the draw of the balls is random and the coin is fair.
   b Calculate the probability of a blue ball and a head.
   c Calculate the probability of a yellow ball and two heads.
   d Calculate the probability that you will not get heads at all.

Section 3: Conditional probability
HOMEWORK 24E

1 A bag contains two 5 p and five 10 p coins. You are asked to draw a coin from the bag at random without replacing any until you get a 10 p coin.
   a Draw a probability tree to show the possible outcomes.
   b Label the branches to show the probability of each event.
   c Calculate the probability of getting a 10 p on your first draw.

   d What is the probability of drawing the two 5 p coins before you draw a 10 p coin?
   e If you draw two 5 p coins on your first two draws, what is the probability of getting a 10 p coin on your third draw? Why?

2 Use the Venn diagram to find:
   a P(A).
   b P(A and B).
   c P(B given that A has occurred).
   d P(A given that B has occurred).

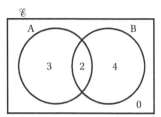

3 A weekend newspaper reports that adults in the UK have an 80% chance of living to be at least 70 years old and a 50% chance of living to be at least 80 years old. What is the conditional probability that an adult that has just turned 70 will live to be 80?

4 A survey of a suburb in Edinburgh shows that 30% of households have no car, half of households have one car, 15% have two cars and 5% have more than two cars.
   a What is the probability that a household has two or more cars?
   b You observe that a household has one car. What is the probability that the same household has at least two cars?

Chapter 24 review

1 Toni is playing a game where she rolls a normal six-sided dice and tosses a coin. If the score on the dice is even, she tosses the coin once. If the score on the dice is odd, she tosses the coin twice.
   a Draw a tree diagram to show all possible outcomes.
   b Assuming the dice and the coin are fair and all outcomes are equally likely, label the branches with the correct probabilities.
   c What is the probability of obtaining two tails?
   d What is the probability of obtaining a five, a head and a tail (in any order)?

2 Two students are to be chosen for a debating team from Amy, James, Nanna, Kenny and Tamara. The teacher choses the team by drawing two names from a hat.
   a How many possible outcomes are there?

b What is the probability that:
  i two girls will be chosen.
  ii two boys will be chosen.
  iii James will be on the team.
  iv the team will have one boy and one girl?

③ An octagonal fair dice is numbered from 1 to 8.
  a Draw a tree diagram to show the probability of getting a multiple of three (M) on two successive rolls of the dice.
  b Use your tree diagram to determine:
   i P(M,M).
   ii P(not M, not M).
   iii P(M, not M).

④ The tree diagram opposite shows the possible outcomes when two coins are tossed. 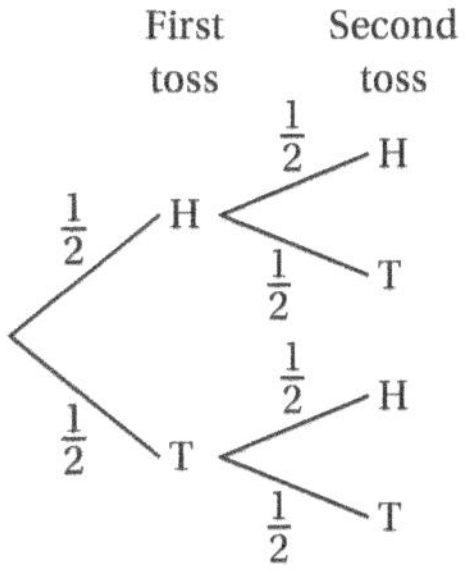
  a Copy and complete the tree diagram to show the possible outcomes when a third coin is tossed.
  b Calculate the probability of tossing three heads.

First toss   Second toss

$\frac{1}{2}$ H $\left\langle \begin{array}{l} \frac{1}{2}\ H \\ \frac{1}{2}\ T \end{array} \right.$

$\frac{1}{2}$ T $\left\langle \begin{array}{l} \frac{1}{2}\ H \\ \frac{1}{2}\ T \end{array} \right.$

c Calculate the probability of getting at least two tails.
d Calculate the probability of getting fewer heads than tails.
e Calculate the probability of getting an equal number of heads and tails.

⑤ Mrs Khan has a choice of ten mobile phone packages. Six of the packages offer free data bundles, five offer a free hands-free kit and three offer both.
  a Draw a Venn diagram to show the given information.
  b Determine:
   i P(getting free data only).
   ii P(getting free data and a free hands-free kit).
   iii P(getting free data or a hands-free kit).

⑥ Two dice are rolled one after another. Determine the probability that:
  a the first dice shows 3, given that the sum of the two scores is 7.
  b the first dice shows 3, given that the sum of the two scores is 5.

25 Powers and roots

Section 1: Index notation
HOMEWORK 25A

💡 **Tip**

Learn the laws of indices.

① Evaluate each expression without using a calculator.
  a 3^3 **b** 8^2 **c** 5^3
  d $3^2 + 2^3$ **e** $4^3 + 2^2$ **f** $5^2 - 2^2$
  g $4^2 \times 5^2$ **h** $2^5 \div 2^1$ **i** $3^5 \div 3^2$

② Use your calculator to evaluate the following.
  a 5^6 **b** 13^4 **c** 12^3
  d 10^5 **e** 24^2 **f** 20^3

💡 **Tip**

Remember, the order of operations says you must calculate the power first.

③ Use a calculator to find the value of each expression.
  a $13^3 - 3^5$ **b** $14^3 \times 13^2$
  c $3^7 + 3^5$ **d** $6^6 + 2^5$
  e $24^3 \div 6^3$

④ Fill in < or > to make each statement true.
  a $5^6 \square 6^5$ **b** $10^4 \square 4^{10}$
  c $11^2 \square 2^{11}$ **d** $11^{10} \square 10^{11}$
  e $5^{10} \square 10^5$

HOMEWORK 25B

1 Write each of the following using positive indices only.

a 3^{-1} **b** 6^{-1} **c** 7^{-1} **d** 5^{-2}

e 2^{-3} **f** 2^{-5} **g** 3^{-5} **h** 7^{-6}

i 24^{-3} **j** x^{-4} **k** n^{-3} **l** y^{-2}

> **Tip**
>
> Remember $4^{-1} = \dfrac{1}{4}$.

2 Express the following with negative indices.

a $\dfrac{1}{2}$ **b** $\dfrac{1}{8}$ **c** $\dfrac{1}{2^2}$ **d** $\dfrac{1}{4^3}$

e $\dfrac{1}{2^4}$ **f** $\dfrac{1}{9^5}$ **g** $\dfrac{1}{7^3}$ **h** $\dfrac{1}{11^4}$

k $\dfrac{1}{y^3}$ **l** $\dfrac{1}{x^4}$ **m** $\dfrac{1}{t^2}$

3 Fill in $=$ or $\neq$ in each of these statements.

a $10^{-2} \,\square\, \dfrac{1}{10^2}$ **b** $7^0 \,\square\, 1$

c $10^{-2} \,\square\, \dfrac{2}{10}$ **d** $6^{-3} \,\square\, \dfrac{1}{6^3}$

e $8^{-3} \,\square\, \dfrac{3}{8}$ **f** $\dfrac{1}{11^5} \,\square\, 11^{-5}$

h $x^{-3} \,\square\, \dfrac{3}{x}$ **i** $\dfrac{1}{y^5} \,\square\, y^{-5}$

> **Tip**
>
> Remember $\neq$ means 'not equal to'.

4 Write each of the following in index notation with base numeral 2. (In other words, rewrite them in the form of 2^x.)

a 1 **b** 8 **c** 32 **d** $\dfrac{1}{4}$

e 0.5 **f** $\sqrt{8}$ **g** $-\sqrt{32}$

Section 2: The laws of indices

HOMEWORK 25C

1 Simplify. Leave the answers in index notation.

a $2^3 \times 2^5$ **b** $10^6 \times 10^3$ **c** $3^4 \times 3^6$

d $4^3 \times 4^{-5}$ **e** $2^{-3} \times 2^7$ **f** $8^0 \times 3^4$

g $3 \times 3^2 \times 3^{-6}$ **h** $4^3 \times 4^2 \times 4$

i $10^4 \times 10^{-6} \times 10^2$

> **Tip**
>
> The number 4^3 is in index notation.

2 Simplify. Leave the answers in index notation.

a $7^4 \div 7^2$ **b** $10^5 \div 10^3$ **c** $10^6 \div 10^2$

d $4^{10} \div 4^0$ **e** $5^6 \div 5$ **f** $10^6 \div 10^6$

g $\dfrac{5^3}{5^{-3}}$ **h** $\dfrac{10^7}{10^{-3}}$ **i** $\dfrac{3^{-4}}{3^{-5}}$ **j** $\dfrac{3^0}{3^4}$

3 Simplify each expression. Give the answers in index notation.

a $(4^3)^3$ **b** $(3^2)^3$ **c** $(5^5)^2$ **d** $(9^3)^2$

e $(5^4)^{-3}$ **f** $(8^{-2})^2$ **g** $(10^2)^{-3}$ **h** $(7^5)^{-2}$

i $(8^4)^0$ **j** $(3^3 \times 3^4)^2$

4 Say whether each statement is true or false. If it is false, write the correct answer.

a $4^2 \times 4^5 = 4^7$ **b** $5^6 \div 5^2 = 5^3$

c $10^9 \div 10^3 = 10^6$ **d** $(7^3)^2 = 7^5$

e $19^0 = 1$ **f** $(5^2)^0 = 5$

HOMEWORK 25D

1 Rewrite each expression using root signs.

a $5^{\frac{1}{2}}$ **b** $6^{\frac{1}{3}}$ **c** $3^{\frac{1}{9}}$

d $7^{\frac{2}{3}}$ **e** $6^{\frac{5}{8}}$ **f** $7^{\frac{4}{9}}$

2 Write in index notation.

a $\sqrt{8}$ **b** $\sqrt[3]{5}$ **c** $\left(\sqrt[3]{5}\right)^5$ **d** $\sqrt[6]{5}$ **e** $\left(\sqrt[3]{8}\right)^4$

3 Evaluate.

a $27^{\frac{1}{3}}$ **b** $64^{\frac{1}{6}}$ **c** $81^{\frac{3}{4}}$ **d** $81^{-\frac{1}{4}}$

Section 3: Working with powers and roots

HOMEWORK 25E

1 Estimate the following roots. Show your working.

a $\sqrt{82}$ **b** $\sqrt[3]{69}$ **c** $\sqrt[4]{53}$

d $\sqrt{37}$ **e** $\sqrt[3]{-150}$ **f** $\sqrt{7}$

2 Use a calculator to find the roots in question 1 correct to two decimal places. How good were your estimates?

3 Find the value of x in each of the following by trial and improvement. Show your working.

a $2^x = 41$ **b** $3^x = 45$ **c** $2^x = 350$

d $x^4 = 2401$ **e** $x^3 = 21$ **f** $x^3 = 5500$

④ Find four different pairs of positive whole values for a and m which will satisfy the equation $a^m = 64$.

⑤ Use trial and improvement methods to find the length of sides of a square with an area of $120\,\text{mm}^2$. Show all your working.

⑥ A cube has a volume of $700\,\text{mm}^3$. Find the lengths of the side of the cube correct to two decimal places using trial and improvement methods. Show all your working.

HOMEWORK 25F

① Alex has £3500. She wants to invest it for ten years in an account that offers 4% growth.
Value of future investment
= Original amount $\times (1.04)^{10}$
a Work out how much money Alex will have in the investment after ten years.
b How much will she have if she decides to spend £1000 and put the rest of the money into this investment?

② Pierre took a mortgage of £75 000 to buy a flat. The bank manager showed him this formula for working out how much he will repay over a 25-year period.
Total amount paid = Mortgage amount $\times (1.05)^{25}$
a Work out how much his mortgage will cost if he takes 25 years to repay it.
b The power of 25 in the formula represents the number of years over which it is repaid. Work out what the total amount paid would be if Pierre paid his mortgage off in 20 years.
c How much would he save by paying over the shorter period?

③ The time (t seconds) a ball takes to hit the ground after being dropped from a height (h in metres) can be found using the formula
$$t = \sqrt{\left(\frac{h}{4.9}\right)}.$$
a Work out how long it will take a ball dropped from a height of 4.8 m to hit the ground.
b Mo drops a ball from a height of 3.5 m and Kai drops a ball from a height of 4.8 m.
i Whose ball will hit the ground first?
ii How many seconds later will the second ball hit the ground?

④ Many measurements in humans and other mammals are in proportion to the mass of the body. Biologists have developed a number of formulae for estimating a number of different measurements. Here are four different formulae which use mass of the body (m) in kilograms to determine other measurements.

Mass of the brain (B) in kilograms	$B = 0.01\,m^{\frac{2}{3}}$
Surface area of the skin (S) in square metres	$S = 0.0096\,m^{\frac{7}{10}}$
Resting metabolic rate (C) (calories consumed at rest)	$C = 70\left(\sqrt[4]{m}\right)^3$
Time (T) it takes for the blood to circulate in seconds	$T = 17.4\left(\sqrt[4]{m}\right)$

a Work out all of these values based on a person with a mass of 74 kg.
b Compare the circulation time for an hippopotamus (average mass 1500 kg) and a human male (average mass 70 kg).
c What is the brain of a 27.5 kg dog likely to weigh?
d Find the surface area of the skin of a rat of mass 0.5 kg and a dog of mass 14.5 kg.
e How many calories does a 125 kg tiger consume while lying in the sun sleeping?

Chapter 25 review

① Write each number in index form.
a $6 \times 6 \times 6 \times 6 \times 6$ b five cubed
c eight squared
d seventeen to the power of six
e $\dfrac{1}{27}$ f $(\sqrt{3})^2$

② Put these expressions in order from smallest to greatest.
a $4^4, \sqrt{64}, 12^2, 3^3, 5 \times \sqrt{144}$
b $6^5, 5^6, 10^4, 92^0, 5^2, 15^2$

③ Write these numbers with positive indices.
a 6^{-4} b 4^{-11} c 7^{-3}

④ Use the laws of indices to simplify each expression and write it as a single power of 6.
a $6^3 \times 6^2$ b $6^5 \times 6^{-3}$ c $6^8 \div 6^3$
d $6^3 \div 6^6$ e $(6^2)^3$ f $(6^{-3})^3$

⑤ Evaluate. Check your answers with a calculator.
a $\sqrt{169}$ b $\sqrt{0.16}$ c $\sqrt[3]{64}$ d $\sqrt[4]{16}$
e $\sqrt[4]{625}$ f $\sqrt{\dfrac{1}{16}}$ g $16^{\frac{3}{2}}$ h $(^-64)^{\frac{2}{3}}$

6 Estimate the length of each side of a cube of volume 42 cm³. Show your working.

7 Find the length of each side of a cube of volume 8000 cm³.

8 The length of time (T seconds) it takes for a pendulum of length (L) in metres to swing through one complete movement can be calculated using the formula $T = 2\pi\left(\dfrac{L}{9.8}\right)^{\frac{1}{2}}$.

Work out how long it will take for a pendulum of length 0.45 m to complete one swing.

9 The radius of a cylindrical container can be found using the formula $r = \sqrt{\left(\dfrac{V}{\pi L}\right)}$, where V is the volume and h is the height of the container. Jasmine has a 15 cm tall cylindrical tin with a volume of 600 cm³. Will she be able to store round biscuits of diameter 9 cm?

26 Standard form

Section 1: Expressing numbers in standard form

HOMEWORK 26A

1 Express each of the following in standard form.

   a 425 000   **b** 45 000
   c 5 020 000   **d** 0.06
   e 0.0002   **f** 511 000
   g 0.000 001 542   **h** 0.000 000 026 52
   i 0.058

Tip

The index number tells you how far and in which direction the number needs adjusting.

2 Express each of the following real life quantities in standard form.

   a In 2014 the population of China was estimated to be 1 366 000 000.
   b The distance from the Earth to the Sun is approximately 149 600 000 km.
   c The Earth is thought to be 4 600 000 000 years old.
   d A hydrogen atom has a radius of 0.00 000 000 001 m.
   e The density of the core of the Sun in 150 000 kg/m
   f The wavelength of violet light is 0.0000004 m.

3 Express each of the following as an ordinary number.

   a 3.6×10^3 **b** 6.2×10^5 **c** 7.9×10^2
   d 6.215×10^5 **e** 3.05×10^{-4} **f** 1.28×10^{-5}
   g 5×10^{-8}

4 Write each quantity out in full as an ordinary number.

   a The Earth orbits the Sun at 2.98×10^4 metres per second.
   b The Moon is 3.84×10^8 away from the Earth.
   c The Sun is 1.5×10^{11} metres away from the Earth on average.
   d There are thought to be 8×10^{10} stars in the Milky Way galaxy.
   e The smallest observable and measurable object is currently 1×10^{-18} m.
   f The mass of an atom of Plutonium-239 is 6.645×10^{-27} g.

Section 2: Calculators and standard form

HOMEWORK 26B

Tip

Make sure you fully understand how your calculator represents standard form.

1 Enter each of these numbers into your calculator using the correct function key and write down what appears on the display.
- **a** 6.3×10^{11}
- **b** 1.9×10^{-6}
- **c** 5.7×10^7
- **d** 1.94×10^{-3}
- **e** 1.52×10^{-10}
- **f** 4.86×10^6
- **g** 3.309×10^{-7}
- **h** 3.081×10^6

2 Here are six different calculator displays giving answers in exponential form. Write each answer correctly in standard form.

a **b** **c**

d **e** **f**

HOMEWORK 26C

1 Use your calculator to do these calculations. Give your answers in standard form correct to three significant figures.
- **a** 4582^7
- **b** $(0.000\,03)^5$
- **c** $0.0008 \div 1200^5$
- **d** $76\,000\,000 \div 0.000\,007$
- **e** $(0.0036)^4 \times (0.00275)^7$
- **f** $(56 \times 274)^3$
- **g** $\dfrac{4489 \times 8630}{0.00006}$
- **h** $\dfrac{7300}{0.0002^5}$
- **i** $\sqrt{7.49} \times 10^6$
- **j** $\sqrt[3]{8.1 \times 10^{-11}}$

2 Work out the following using your calculator. Give the answers in standard form correct to five significant figures.
- **a** $(1.009)^5$
- **b** $123\,000\,000 \div 0.00076$
- **c** $(97 \times 876)^4$
- **d** $(0.0098)^4 \times (0.0032)^3$
- **e** $\dfrac{8543 \times 9210}{0.000\,034}$
- **f** $\dfrac{9745}{(0.0004)^4}$
- **g** $\sqrt[3]{(4.2 \times 10^{-8})}$

Section 3: Working in standard form
HOMEWORK 26D

1 Simplify, giving the answers in standard form.
- **a** $(3 \times 10^{11}) \times (5 \times 10^{12})$
- **b** $(6.4 \times 10^9) \times (2 \times 10^5)$
- **c** $(9 \times 10^{15}) \div (3 \times 10^{10})$
- **d** $(2.6 \times 10^7) \div (7 \times 10^3)$
- **e** $(5.8 \times 10^{54}) \div (6 \times 10^{25})$

2 Simplify each of the following giving all answers in standard form.
- **a** $(3 \times 10^{-3}) \times (6 \times 10^{-17})$
- **b** $(1.3 \times 10^{-9}) \times (6 \times 10^{-5})$
- **c** $(1.8 \times 10^{-7}) \times (7.1 \times 10^{-2})$
- **d** $(1.2 \times 10^{-5}) \times (5 \times 10^3)$
- **e** $(8 \times 10^{16}) \div (9.2 \times 10^{-13})$
- **f** $(8 \times 10^{-22}) \div (1 \times 10^{17})$

3 Carry out these calculations without using your calculator. Leave the answer in standard form.
- **a** $(4 \times 10^{13}) \times (5 \times 10^{15})$
- **b** $(5.5 \times 10^7) \times (6 \times 10^4)$
- **c** $(6 \times 10^{10})^3$
- **d** $(1.7 \times 10^{-5}) \times (1.8 \times 10^{-7})$
- **e** $(6 \times 10^{15}) \times (3 \times 10^{10})$
- **f** $(7 \times 10^{15}) \div (8 \times 10^{13})$
- **g** $(1.68 \times 10^9) \div (8 \times 10^6)$
- **h** $(7 \times 10^{-10}) \div (8 \times 10^{-14})$

4 The speed of light is approximately 3×10^8 metres per second. How far will light travel in:
- **a** 15 seconds?
- **b** 30 seconds?
- **c** 10^3 seconds?
- **d** 3×10^5 seconds?

5 There are approximately 1.1×10^{14} cells in each human body.
- **a** How many cells would there be in a class of 30 students? Give the answer in standard form and as an ordinary number.
- **b** If there were 7.2×10^9 people on Earth, how many human cells are there on the planet?

HOMEWORK 26E

1 Carry out these calculations without using a calculator. Give your answers in standard form.
- **a** $(4 \times 10^7) + (2 \times 10^7)$
- **b** $(4 \times 10^{-4}) - (1.5 \times 10^{-4})$
- **c** $(2.5 \times 10^6) + (3 \times 10^7)$
- **d** $(7 \times 10^8) - (4 \times 10^7)$
- **e** $(6 \times 10^{-5}) + (3 \times 10^{-4})$
- **f** $(8 \times 10^{-3}) - (2.5 \times 10^{-5})$

2 Mars has a surface area of approximately $1.45 \times 108\,\text{km}^2$ and the Earth has a surface area of approximately $5.1 \times 10^8\,\text{km}^2$.
- **a** Which planet has the greater surface area?
- **b** How much larger is it?
- **c** Saturn has a surface area of 4.27×10^{10}. How much bigger is the surface area of Saturn than:
  - **i** Mars?
  - **ii** Earth?

3 The Earth is approximately 1.5×10^8 km from the Sun and Mercury is approximately 7.48×10^7 km from the Sun.

a What is the closest distance possible between the two planets?

b What is the maximum possible distance between the two planets?

HOMEWORK 26F

1 Earth is 4.6×10^9 years old.

a The Hadean period of geological time ended about 3.8×10^9 years ago.
How long did this period last?

b Dinosaurs were first on the planet 2.3×10^8 years ago. How many years after the formation of Earth was this?

c The dinosaur era ended 6.5×10^7 years ago. How long did dinosaurs exist on Earth?

2 A nanometer (nm) is a very small unit of measure. One nanometer is 1.0×10^{-9} m. Express the following measurements in nanometers.

a 53 m   b 61 mm

3 Jupiter has a mass of approximately 1.89813×10^{27} kilograms.
The planet Saturn has a mass of approximately 5.683×10^{24} kg.

a Which has the greater mass?

b How many times heavier is the greater mass compared with the smaller mass?

4 Light travels at a speed of 3×10^8 metres per second. The Sun is an average distance of 1.5×10^{11} m from Earth, and Jupiter is an average 6.3×10^{11} m from the Sun.

a Work out how long it takes light from the Sun to reach Saturn (in seconds).
Give your answer in both ordinary numbers and standard form.

b How much longer does it take for the light to reach Saturn than it does Earth?
Give your answer in both ordinary numbers and standard form correct to three significant places.

5 An immunologist cultures two sets of bacteria. Culture A contains 9.2×10^{12} bacterial cells, Culture B contains 5.2×10^8 bacterial cells.

She combines the two cultures in one incubation flask.

a How many cells are there when she combines the two cultures?

b The bacteria numbers double every eight hours. How many cells will there be after two days?

Chapter 26 review

1 Express the following numbers in standard form.

a 65 000   b 50 000 000   c 6 215 000
d 78 000 000 000   e 0.00015   f 0.0009

2 Write the following in ordinary numbers.

a 5.6×10^4 b 2.8×10^6
c 2.475×10^7 d 2.056×10^{-4}
e 7.15×10^{-7} f 6.0×10^{-10}
g 3.8×10^{-6}

3 Use a calculator and give the answers in standard form.

a $5 \times 10^4 + 9 \times 10^6$
b $3.27 \times 10^{-3} \times 2.4 \times 10^2$
c $5(8.1 \times 10^9 - 2 \times 10^7)$
d $(3.2 \times 10^{-1}) - (2.33 \times 10^{-3})$ (to 3 sf)

4 Simplify the following without using a calculator and give the answers in standard form.

a $(5.26 \times 10^7) + (8.2 \times 10^7)$
b $(8.2 \times 10^5) \times (6.3 \times 10^9)$
c $(6 \times 10^4) + (5 \times 10^3)$
d $(3 \times 10^6) \div (2 \times 10^5)$

5 The UK has an approximate area of 2.4×10^5 km². The USA has an area of approximately 9.8×10^6 km².

a What is the difference in the areas of the two countries?
Give the answer in standard form.

b What is the combined area of the two countries?
Give the answer in standard form.

c How many times could the map of the UK fit onto the map of USA if they were drawn at the same scale?

6 The diameter of the planet Venus at its equator is 1.2×10^5 km.

a What is the approximate radius of Venus at its equator?

b What is the approximate circumference of Venus at its equator (remember $C = 2\pi r$)

c Assuming Venus is a sphere, and that the volume of a sphere is $\frac{4}{3}\pi r^3$, calculate the volume of the planet Venus to three significant figures.

27 Surds

Section 1: Approximate and exact values

HOMEWORK 27A

1 Use a calculator to find the approximate value of each expression correct to three decimal places.

  a $3\sqrt{3}$ **b** $2\sqrt{6}$ **c** $^-4\sqrt{3}$
  d $^-6\sqrt{11}$ **e** $12\sqrt{3}$ **f** $5(3\sqrt{2})$
  g $^-5(2\sqrt{27})$ **h** $^-3(4\sqrt{45})$

> ### Tip
> Think of a square root sign like a bracket.

> ### Tip
> Some calculators will try to leave the answer in surd form – press the key that converts to decimal form.

2 Find the approximate value of each expression, giving your answers correct to three decimal places.

  a $\sqrt{4} + \sqrt{3}$ **b** $\sqrt{10} - \sqrt{5}$ **c** $\sqrt{4 + 3}$
  d $\sqrt{11 + 4}$ **e** $\sqrt{14 + 3}$ **f** $\sqrt{9 - 6}$
  g $\sqrt{4 - 3}$ **h** $3\sqrt{3} + 5\sqrt{2}$ **i** $4\sqrt{2} + 5\sqrt{3}$

HOMEWORK 27B

1 What is the exact circumference of a circle with a radius of $\sqrt{3}$ cm?

2 A rectangular sheet of paper has side lengths 10 cm and 15 cm.
What is the length of the diagonal?

3 Two identical square tiles have a combined area of 68 cm².
What is the exact length of the sides of each square?

4 The area of a square field is 5 hectares.
  a What is the exact length of each side of the field?
  b What is the exact diagonal distance across the field?

5 Jamal makes square ceramic tiles to individual designs.
Usually his designs have a diagonal of 6 cm.
  a What is the exact length of each side of such a square tile?
  b Use the exact length to calculate the exact area of a square tile.
  c Give the length of the sides correct to:
    i two decimal places.
    ii three decimal places.
  d Calculate the area of each tile using the approximate values.
  e Jamal needs to know how much clay he will need for 100 tiles. Use both answers from **c** and find the difference between them, calculating to two decimal places and three decimal places

Section 2: Manipulating surds

HOMEWORK 27C

1 Simplify.
  a $\sqrt{28}$ **b** $\sqrt{32}$ **c** $\sqrt{80}$ **d** $\sqrt{112}$
  e $\sqrt{75}$ **f** $\sqrt{300}$ **g** $\sqrt{150}$ **h** $\sqrt{500}$

2 The following surds are already in their simplest form.
$\sqrt{3}\sqrt{7}$ $\sqrt{17}$ $\sqrt{59}$ $\sqrt{89}$
How can you tell?

3 Write each surd in its simplest form.
   a $3\sqrt{12}$ b $^-2\sqrt{18}$ c $4\sqrt{20}$ d $^-3\sqrt{80}$
   e $4\sqrt{54}$ f $^-7\sqrt{112}$ g $^-3\sqrt{63}$ h $^-2\sqrt{48}$
   i $8\sqrt{96}$ j $6\sqrt{200}$

4 Complete the following.
   a $3\sqrt{6} = 3 \times \sqrt{6}$ b $^-5\sqrt{7} = ^-5 \times \sqrt{7}$
   $\quad = \sqrt{} \times \sqrt{6}$ $\quad = ^-\sqrt{} \times \sqrt{7}$
   $\quad = \sqrt{} \times 6$ $\quad = ^-\sqrt{} \times 7$
   $\quad = \sqrt{}$ $\quad = ^-\sqrt{}$

Tip

Remember to leave the negative sign outside the square root symbol.

5 Rewrite the following in the form $\sqrt{n}$.
   a $5\sqrt{3}$ b $6\sqrt{5}$ c $2\sqrt{6}$ d $5\sqrt{12}$
   e $^-3\sqrt{8}$ f $^-2\sqrt{5}$ g $^-8\sqrt{15}$ h $^-3\sqrt{13}$
   i $12\sqrt{7}$ j $14\sqrt{3}$

6 Write these surds in order of size, smallest first.
   $4\sqrt{2}$ $3\sqrt{3}$ $2\sqrt{3}$ $9\sqrt{3}$ $6\sqrt{5}$
   $5\sqrt{6}$ $3\sqrt{8}$ $2\sqrt{12}$ $4\sqrt{2}$

HOMEWORK 27D

1 Simplify by adding or subtracting.
   a $3\sqrt{6} + 3\sqrt{7} + 2\sqrt{6} + 5\sqrt{7}$
   b $11\sqrt{3} + 2\sqrt{5} - 6\sqrt{3} + 4\sqrt{5}$
   c $3\sqrt{7} - 5\sqrt{8} + 3\sqrt{7} + 2\sqrt{8}$
   d $3\sqrt{10} + 3\sqrt{5} - 5\sqrt{10} - 7\sqrt{5}$

2 Simplify.
   a $\sqrt{3} + \sqrt{18}$ b $\sqrt{20} - \sqrt{5}$ c $\sqrt{8} + \sqrt{32}$
   d $2\sqrt{3} - \sqrt{12}$ e $3\sqrt{5} - \sqrt{20}$ f $9\sqrt{45} - 2\sqrt{20}$

3 Simplify.
   a $\sqrt{12} + \sqrt{28} - 2\sqrt{3}$
   b $2\sqrt{32} + 2\sqrt{48} - \sqrt{96} - 2\sqrt{48}$
   c $3\sqrt{125} - 2\sqrt{50} - \sqrt{75}$
   d $2\sqrt{12} - 2\sqrt{24} - \sqrt{63} + 2\sqrt{54}$

4 A rectangular lawn has side dimensions $(2 - \sqrt{5})$ m by $(2 + \sqrt{24})$ m. Calculate the exact perimeter of the lawn.

HOMEWORK 27E

1 Simplify.
   a $\sqrt{7} \times \sqrt{10}$ b $3\sqrt{6} \times 4\sqrt{5}$ c $^-2\sqrt{10} \times 3\sqrt{4}$
   d $3\sqrt{12} \times 3\sqrt{8}$ e $3\sqrt{12} \times 5\sqrt{12}$ f $3\sqrt{5} \times 5\sqrt{3}$

2 Simplify.
   a $\dfrac{\sqrt{5}}{\sqrt{15}}$ b $\dfrac{\sqrt{3}}{\sqrt{21}}$ c $\dfrac{\sqrt{54}}{\sqrt{27}}$

   d $\dfrac{4\sqrt{6}}{4}$ e $\dfrac{7\sqrt{5}}{14}$ f $\sqrt{\dfrac{3\sqrt{72}}{6}}$

   g $\dfrac{8\sqrt{26}}{\sqrt{2}}$ h $\dfrac{16\sqrt{10}}{4\sqrt{2}}$

3 Simplify fully.
   a $3\sqrt{12} \times \dfrac{4\sqrt{6}}{4}$ b $3\sqrt{6} \times \dfrac{4\sqrt{7}}{\sqrt{12}}$

   c $4\sqrt{7} \times \dfrac{3\sqrt{10}}{3\sqrt{20}}$ d $\sqrt{5} \times \dfrac{\sqrt{15}}{3\sqrt{5}}$

   e $4\sqrt{7} \times \dfrac{^-2\sqrt{3}}{3\sqrt{15}}$ f $^-\sqrt{10} \times \dfrac{\sqrt{27}}{3\sqrt{12}} \times 2\sqrt{2}$

4 Expand and simplify.
   a $\sqrt{4}(\sqrt{5} + 2)$ b $2\sqrt{5}(6 - \sqrt{2})$
   c $(\sqrt{3} + 2)(\sqrt{7} - 5\sqrt{2})$ d $(3\sqrt{5} + \sqrt{6})^2$
   e $(\sqrt{3} + 2)(\sqrt{2} + 5)$ f $(3 - \sqrt{3})(\sqrt{6} - 1)$
   g $(3\sqrt{5} - \sqrt{2})(3\sqrt{7} + \sqrt{2})$
   h $(\sqrt{6} + \sqrt{3})^2$ i $(\sqrt{7} - \sqrt{8})^2$

5 Express each of the following in simplest form with a rational denominator.
   a $\dfrac{6}{\sqrt{3}}$ b $\dfrac{2}{\sqrt{5}}$ c $\dfrac{^-1}{\sqrt{3}}$

   d $\dfrac{\sqrt{3}}{\sqrt{2}}$ e $5\sqrt{\dfrac{5}{2}}$ f $\dfrac{^-4}{2\sqrt{6}}$

   g $\dfrac{(3 + \sqrt{2})}{2\sqrt{3}}$ h $\dfrac{(4 + \sqrt{3})}{\sqrt{3}}$

6 Rationalise each denominator and simplify.
   a $\dfrac{4}{(3 - \sqrt{2})}$ b $\dfrac{\sqrt{3}}{(\sqrt{11} + 3)}$ c $\dfrac{\sqrt{5}}{(\sqrt{7} - \sqrt{2})}$

   d $\dfrac{5}{(2\sqrt{3} + 5)}$ e $\dfrac{9}{(\sqrt{6} + 1)}$

Section 3: Working with surds
HOMEWORK 27F

1 Calculate the exact area of a rectangular field that is $(7 + \sqrt{6})$ hectares by $(7 - \sqrt{6})$ hectares.

2 If the area of a pizza is 200 cm², what is its exact radius?

3 A sheet of A4 paper has sides in the ratio $1 : \sqrt{2}$. If a sheet of A4 is 210 mm wide, what would be the length of the diagonal?

4 The dimensions of a cuboid are: $\sqrt{2} + 3, 6 - \sqrt{3}$ and $\sqrt{5}$.
What is the surface area of the cuboid?
Leave your answer in surd form.

5 Given that ABC is a right-angled isosceles triangle with AB = BC, determine cos A, leaving your answer in surd form.

6 A square has a side length of $4\sqrt{6}$.
What is:
a the area of the square?
b the perimeter of the square?

7 Find the exact perimeter of a square whose area is 250 cm².

8 A right-angled triangle has its two shortest sides as $\sqrt{2}$ and $\sqrt{3}$.
Calculate the perimeter and area of the triangle.

Chapter 27 review

1 Which of the following answers is a correct simplification?
If neither is correct, find the correct one.

	Question	Answer A	Answer B
1	$\sqrt{54}$	$\sqrt{3} \times \sqrt{18}$	$3\sqrt{6}$
2	$\sqrt{96}$	$4\sqrt{6}$	$6\sqrt{4}$
3	$2\sqrt{4} + 2\sqrt{3} + 4\sqrt{4} - \sqrt{3}$	$12\sqrt{3}$	13.73
4	$\sqrt{18} + \sqrt{6}$	$5\sqrt{2}$	$\sqrt{24}$
5	$\sqrt{6} \times \sqrt{6} \times \sqrt{6} \times \sqrt{6}$	36	$4\sqrt{6}$
6	$\sqrt{5} + \sqrt{5} \times \sqrt{5} + \sqrt{5}$	$4\sqrt{5}$	$2\sqrt{5}$
7	$\dfrac{6\sqrt{45}}{2\sqrt{6}}$	$\dfrac{3\sqrt{30}}{2}$	$3\sqrt{15}$
8	$\dfrac{(2 + \sqrt{7})}{\sqrt{7}}$	$\sqrt{7} + 2$	$\dfrac{(7 + 2\sqrt{7})}{7}$
9	$\dfrac{(1 + 5\sqrt{6})}{5}$	$1 + \sqrt{6}$	$1 - \sqrt{6}$
10	$\dfrac{3}{(4\sqrt{6} + 3)}$	$\dfrac{1}{7\sqrt{6}}$	$\dfrac{1}{4}$

28 Plane vector geometry

Section 1: Vector notation and representation
HOMEWORK 28A

Tip

In vectors the horizontal direction is given by the top figure, the vertical by the bottom one.

1 The diagram opposite shows eight vectors.
Use vector notation to write down each vector.
The first one has been done for you.

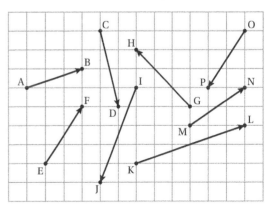

$$\overrightarrow{AB} = \begin{pmatrix} 3 \\ 1 \end{pmatrix}$$

2 Draw a pair of axes where x and y vary from $^-10$ to 10.

Plot the point $A(1, 2)$.

Plot the points B, C, D, E, F and G where:

$$\overrightarrow{AB} = \begin{pmatrix} 4 \\ -2 \end{pmatrix} \quad \overrightarrow{BC} = \begin{pmatrix} -1 \\ 9 \end{pmatrix} \quad \overrightarrow{CD} = \begin{pmatrix} -7 \\ -10 \end{pmatrix}$$

$$\overrightarrow{DE} = \begin{pmatrix} 11 \\ -1 \end{pmatrix} \quad \overrightarrow{EF} = \begin{pmatrix} 2 \\ 10 \end{pmatrix} \quad \overrightarrow{FG} = \begin{pmatrix} -16 \\ -11 \end{pmatrix}$$

3 **a** Find the vector from point P with coordinates $(2, 6)$ to the point Q with coordinates $(^-3, 5)$.

b What is the vector $\overrightarrow{QP}$?

4 **a** Find the vector from point A with coordinates $(^-3, 4)$ to the point B with coordinates $(^-7, ^-2)$.

b Use your answer to find the coordinates of the midpoint of AB.

5 The following vectors describe how to move between points P, Q, R and S.

$$\overrightarrow{PQ} = \begin{pmatrix} -3 \\ 5 \end{pmatrix} \quad \overrightarrow{QR} = \begin{pmatrix} 3 \\ 0 \end{pmatrix} \quad \overrightarrow{RP} = \begin{pmatrix} 0 \\ -5 \end{pmatrix} \quad \overrightarrow{QS} = \begin{pmatrix} 6 \\ -5 \end{pmatrix}$$

a Draw a diagram to show how the points are positioned to form the quadrilateral PQRS.

b What shape is PQRS?

6 The vector $\begin{pmatrix} 9 \\ -7 \end{pmatrix}$ describes the displacement from point A to point B.

a What is the vector from point B to point A?

b Point A has coordinates $(^-4, 3)$. What are the coordinates of point B?

7 Calculate the magnitude of the following vectors.

a $\begin{pmatrix} 3 \\ 4 \end{pmatrix}$ **b** $\begin{pmatrix} 4 \\ 6 \end{pmatrix}$ **c** $\begin{pmatrix} -2 \\ 5 \end{pmatrix}$ **d** $\begin{pmatrix} 4 \\ -5 \end{pmatrix}$ **e** $\begin{pmatrix} -3 \\ -4 \end{pmatrix}$

Section 2: Vector arithmetic
HOMEWORK 28B

1 $\mathbf{p} = \begin{pmatrix} -3 \\ 4 \end{pmatrix} \quad \mathbf{q} = \begin{pmatrix} 3 \\ -2 \end{pmatrix} \quad \mathbf{r} = \begin{pmatrix} 7 \\ -3 \end{pmatrix} \quad \mathbf{s} = \begin{pmatrix} -9 \\ -7 \end{pmatrix}$

Write each of these as a single vector.

a $\mathbf{p} + \mathbf{q}$ **b** $\mathbf{q} - \mathbf{r}$

c $\mathbf{s} - \mathbf{r}$ **d** $\mathbf{q} + \mathbf{s}$

e $3\mathbf{p}$ **f** $^-4\mathbf{r}$

g $\mathbf{p} + \mathbf{q} + \mathbf{r}$ **h** $3\mathbf{q} - \mathbf{r}$

i $2\mathbf{r} + 3\mathbf{s}$ **j** $2\mathbf{p} + 3\mathbf{r} - \mathbf{s}$

2 Write down three vectors which are parallel to $\begin{pmatrix} -2 \\ 4 \end{pmatrix}$.

3 Find the values of x, y and z in these vector calculations.

a $\begin{pmatrix} x \\ -3 \end{pmatrix} + \begin{pmatrix} 3 \\ -7 \end{pmatrix} = \begin{pmatrix} -1 \\ y \end{pmatrix}$ **b** $\begin{pmatrix} 6 \\ x \end{pmatrix} - \begin{pmatrix} y \\ -2 \end{pmatrix} = \begin{pmatrix} 8 \\ -5 \end{pmatrix}$

c $\begin{pmatrix} x \\ -5 \end{pmatrix} + \begin{pmatrix} 5 \\ y \end{pmatrix} = \begin{pmatrix} 0 \\ 0 \end{pmatrix}$ **d** $\begin{pmatrix} 4 \\ -2 \end{pmatrix} = x\begin{pmatrix} 12 \\ -6 \end{pmatrix}$

e $x\begin{pmatrix} -4 \\ -3 \end{pmatrix} + y\begin{pmatrix} 2 \\ 3 \end{pmatrix} = \begin{pmatrix} -16 \\ -15 \end{pmatrix}$

4 In the diagram $\overrightarrow{AB} = \begin{pmatrix} 18 \\ 12 \end{pmatrix}$

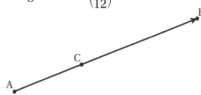

The ratio of $AC : CB$ is $1 : 2$.

a Find $\overrightarrow{AC}$. **b** Find $\overrightarrow{CB}$.

5 These vectors describe how to move between the points P, Q, R and S, which are four sides of a quadrilateral.

$$\overrightarrow{PQ} = \begin{pmatrix} 4 \\ 1 \end{pmatrix} \quad \overrightarrow{QS} = \begin{pmatrix} -1 \\ -5 \end{pmatrix} \quad \overrightarrow{PR} = \begin{pmatrix} 7 \\ -3 \end{pmatrix}$$

a What can you say about sides PQ and SR?

b What type of quadrilateral is PQRS?

6 ABCD is a quadrilateral.

If the vectors $\overrightarrow{AD}$ and $\overrightarrow{BC}$ are parallel and of equal length, what shape could ABCD be?

Section 3: Using vectors in geometric proofs
HOMEWORK 28C

1 In the diagram below $\overrightarrow{AB} = \begin{pmatrix} 8 \\ 3 \end{pmatrix}$ and $\overrightarrow{BC} = \begin{pmatrix} 2 \\ -4 \end{pmatrix}$.

M is the midpoint of AB.

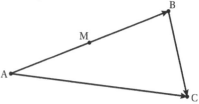

Find:

a $\overrightarrow{BA}$. **b** $\overrightarrow{BA} + \overrightarrow{AC}$. **c** $\overrightarrow{AM}$.

2 Two triangles ABC and DEF have the vertices
A = (1, 4), B = (5, 3), C = (6, 7), D = (5, 2),
E = (9, 1) and F = (10, 5)
Compare the vectors:

a $\overrightarrow{AB}$ and $\overrightarrow{DE}$.

b $\overrightarrow{AC}$ and $\overrightarrow{DF}$.

c What must be true about the triangles ABC and DEF?

3 PQRS is a parallelogram.

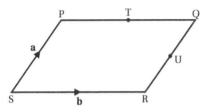

T is the midpoint of PQ, and U is the midpoint of QR.
$\overrightarrow{SP}$ = **a** and $\overrightarrow{SR}$ = **b**
Find the following vectors.
Explain your answers.

a $\overrightarrow{PT}$ b $\overrightarrow{QU}$ c $\overrightarrow{RT}$

d $\overrightarrow{PU}$ e $\overrightarrow{TS}$ f $\overrightarrow{UT}$

4 The diagram below shows a tessellating pattern of congruent quadrilaterals and the vectors **p**, **q** and **n**.

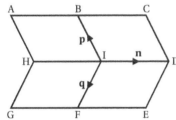

Find the vectors.

a $\overrightarrow{BA}$ b $\overrightarrow{BD}$ c $\overrightarrow{HE}$ d $\overrightarrow{FC}$
e $\overrightarrow{CG}$ f $\overrightarrow{EA}$ g $\overrightarrow{DA}$

5 Triangles ABC and AED are similar.
The ratio CD : CD = 1 : 3.
Find the vectors.

a $\overrightarrow{CA}$
b $\overrightarrow{DE}$
c $\overrightarrow{AE}$
d $\overrightarrow{BA}$

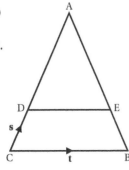

6 The diagram below shows a regular hexagon ABCDEF. H is the centre of the hexagon.

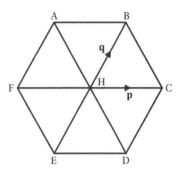

Find the vectors.

a $\overrightarrow{BA}$ b $\overrightarrow{DA}$ c $\overrightarrow{FD}$ d $\overrightarrow{CE}$ e $\overrightarrow{DB}$

7 In a wind tunnel the air is blowing at 6 metres per second.
The wind tunnel is 1 m wide.
A projectile is fired across the tunnel at 0.2 metres per second.

a How far away from perpendicular is the projectile when it reaches the other side of the tunnel?

b How far does the projectile travel?

8 On a chessboard there are 64 squares.
A knight can move two squares horizontally and one square vertically or two squares vertically and one square horizontally.
Assuming there is nothing in the way, what is the smallest number of moves that a knight can take to get from one corner square to the opposite diagonal square?

Chapter 28 review

1 Explain why (3, 4) is different from $\begin{pmatrix} 3 \\ 4 \end{pmatrix}$.

2 Which of the following vectors are parallel?

a $\begin{pmatrix} 3 \\ 4 \end{pmatrix}$ b $\begin{pmatrix} 8 \\ 9 \end{pmatrix}$ c $\begin{pmatrix} 6 \\ -6 \end{pmatrix}$

d $\begin{pmatrix} 4 \\ 3 \end{pmatrix}$ e $\begin{pmatrix} 9 \\ 12 \end{pmatrix}$ f $\begin{pmatrix} 2 \\ -2 \end{pmatrix}$

3 Calculate.

a $\begin{pmatrix} 4 \\ -5 \end{pmatrix} + \begin{pmatrix} 5 \\ 7 \end{pmatrix}$ b $\begin{pmatrix} 6 \\ 3 \end{pmatrix} - \begin{pmatrix} 5 \\ -2 \end{pmatrix}$ c $4 \begin{pmatrix} 1 \\ -6 \end{pmatrix}$

4 In the diagram, M is the midpoint of BC.

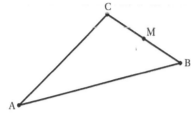

A is the point (1, 1), B is the point (7, 5) and C is the point (5, 7).

Find:

a $\overrightarrow{AC}$. **b** $\overrightarrow{CM}$. **c** $\overrightarrow{AM}$.

5 The diagram below shows the triangle JKL.

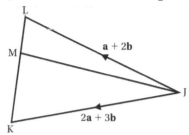

KM : ML = 3 : 1

Find:

a $\overrightarrow{LK}$ **b** $\overrightarrow{ML}$ **c** $\overrightarrow{JM}$

29 Plane isometric transformations

Section 1: Reflections

HOMEWORK 29A

1 Reflect the triangle in the line $x = 2$ and the resultant image in the line $y = {}^-2$.

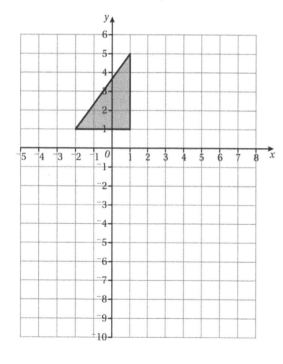

2 Reflect the shape in the line $y = x$ and the resultant image in the line $y = {}^-x$

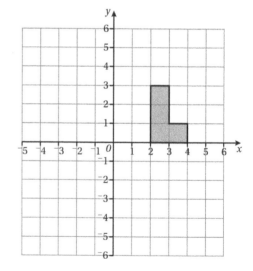

3 Carry out the following reflections.

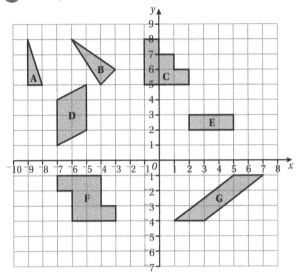

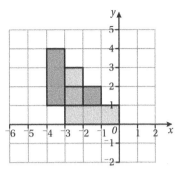

b

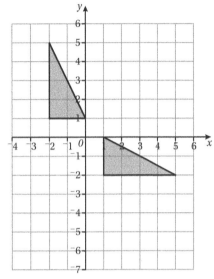

c

a Shape A in the line $x = {}^-6$.
b Shape B in the line $y = 3$.
c Shape C in the line $y = x$.
d Shape D in the line $x = {}^-5$.
e Shape E in the line $y = x$.
f Shape F in the line $y = {}^-2$.
g Shape G in the line $y = {}^-x$.

HOMEWORK 29B

1 Find the equation of the mirror line in each of the following reflections.

a

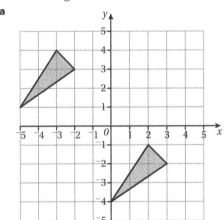

2 Describe the reflection that takes:

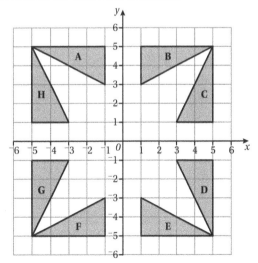

a shape G to D.
b shape F to A.
c shape C to H.
d shape B to C.
e shape E to D.
f shape A to H.

95

3 Trace each pair of shapes and construct the mirror line for the reflection.

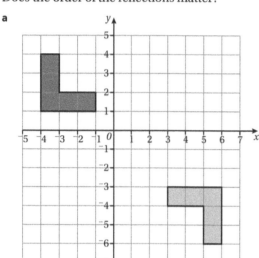

a **b** **c**

4 In each diagram below describe the multiple reflections that take the original dark-shaded shape to its image.
Is there more than one answer?
Does the order of the reflections matter?

a

b

Section 2: Translations
HOMEWORK 29C

1 Translate the shape using the given vectors.

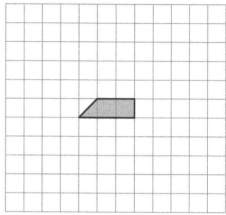

a $\begin{pmatrix} 4 \\ 2 \end{pmatrix}$ **b** $\begin{pmatrix} 3 \\ -2 \end{pmatrix}$

c $\begin{pmatrix} -4 \\ -5 \end{pmatrix}$ **d** $\begin{pmatrix} -4 \\ -1 \end{pmatrix}$

2 Translate each shape by the vector given.

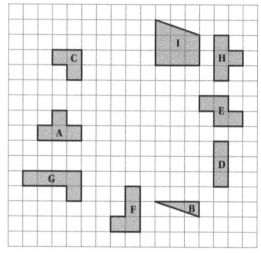

a A $\begin{pmatrix} 5 \\ -2 \end{pmatrix}$ **b** B $\begin{pmatrix} -3 \\ 7 \end{pmatrix}$

c C $\begin{pmatrix} 7 \\ -5 \end{pmatrix}$ **d** D $\begin{pmatrix} -8 \\ 4 \end{pmatrix}$

e E $\begin{pmatrix} -4 \\ -3 \end{pmatrix}$ **f** F $\begin{pmatrix} 3 \\ 7 \end{pmatrix}$

g G $\begin{pmatrix} 6 \\ 6 \end{pmatrix}$ **h** H $\begin{pmatrix} -8 \\ -6 \end{pmatrix}$

i I $\begin{pmatrix} -3 \\ -5 \end{pmatrix}$

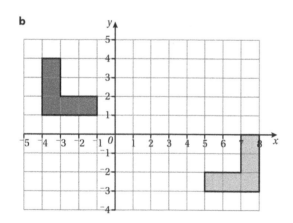

HOMEWORK 29D

1 The complete square below has been made by fitting the pieces A to I together as shown.

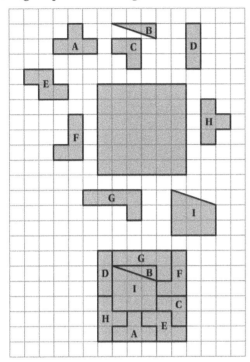

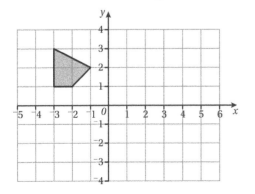

Write down the vector that will translate each separate piece A to I to the corresponding part of the empty grey square.

Section 3: Rotations
HOMEWORK 29E

1 Rotate each shape as directed.
   a 90° clockwise about the origin.

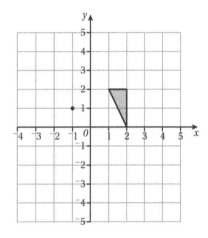

b 90° anticlockwise about the point (1, 1).

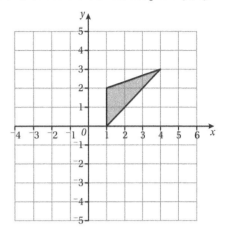

2 Rotate each shape as directed about the marked centre.
   a 90° clockwise

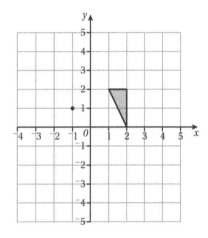

b 90° anticlockwise

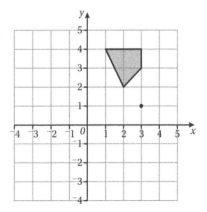

3 Rotate the arrow shape 90°, 180° and 270° clockwise about the origin.

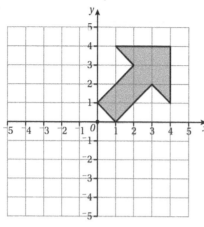

4 Rotate the shape 90°, 180° and 270° clockwise about the point (⁻1, 1).

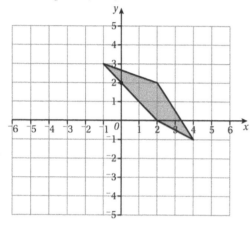

5 The following image was designed by drawing a triangle and rotating it around the point (1, 1) in multiples of 90°. Write down the coordinates of each of the marked points.

What would be the coordinates if shape A was rotated in multiples of 90° around the point (2, 2)?

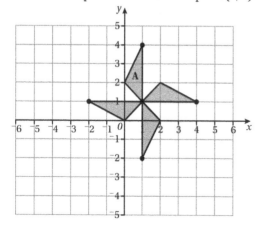

HOMEWORK 29F

1 Describe each of the following rotations.

a

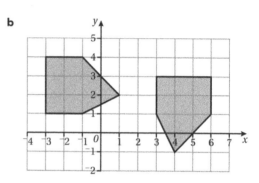

b

2 The diagram below shows part of a crossword grid.

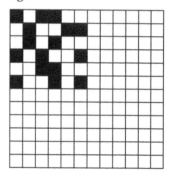

Copy the diagram on squared paper.
Label the bottom left corner the origin, and label the axes.
Rotate the square formed by (0, 6), (6, 6), (6, 12) and (0, 12) 90°, 180° and 270° about the point (6, 6) and complete the grid.
How many white squares are there?

3 Use construction to locate the centres of rotation for each pair of shapes.

Section 4: Combined transformations
HOMEWORK 29G

1 Make a copy of this diagram. Translate shape B through a vector of $\begin{pmatrix} -2 \\ 3 \end{pmatrix}$ and label it B′. Rotate B′ 180° about the point ($^-$2, 6) and label it B″. Describe the single transformation that maps B to B″.

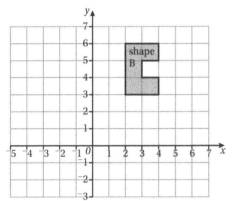

2 Make a copy of this diagram. Reflect shape C in the x axis. Label this shape C′. Rotate C′ 90° anticlockwise about the point ($^-$1, 2) and label it C″. Reflect C″ in the line $y = 3$. Describe the single transformation that maps shape C onto C‴.

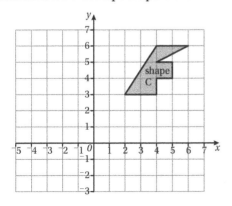

Chapter 29 review

1 Which of the following statements are true? Give reasons, including diagrams if necessary.
   a Any reflected object is congruent to its image.
   b A reflection about the x axis is the same as a rotation of 180°.
   c A triangle ABC is such that A is (5, 3), B is (4, 2) and C is (6, 6) is rotated about the origin 180°. The coordinates of the image are A′ = ($^-$5, $^-$6), B′ = ($^-$5, $^-$3), C′ = ($^-$2, $^-$4).

2 The triangle XYZ has vertices X = (6, 3), Y = (3, 2) and Z = (4, 6). Write down the coordinates of the images of X, Y and Z after the following transformations.
   a Translation by the vector $\begin{pmatrix} 5 \\ 2 \end{pmatrix}$.
   b Rotation of 90° clockwise in the point (2, 2).
   c Reflection in the line $y = 2$.

3 Make a copy of this diagram.

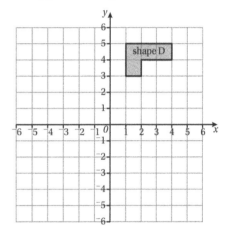

Rotate the shape D 90° clockwise about the origin.
Reflect the resultant shape D′ in the line $x = 0$ to produce D″.
What single transformation would map D″ onto D?

30 Congruent triangles

Section 1: Congruent triangles
HOMEWORK 30A

Tip

Congruent means exactly the same shape and size.

1 Which of the following triangles are congruent?
(All lengths are in centimetres.)

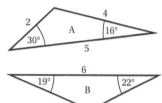

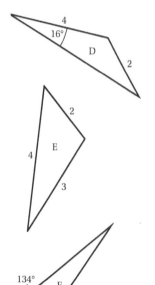

2 Which of the following pairs of triangles are congruent? Justify your answer.
(All lengths are in centimetres.)

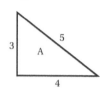 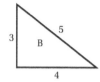

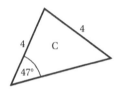 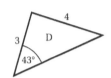

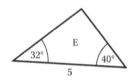 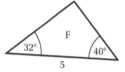

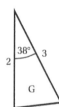 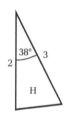

3 **a** Are triangles ABC and DEC congruent?

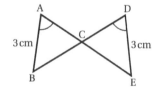

b Explain how you know.
c What must be true of the point C?

4 Explain why triangles FGI and HGI are congruent.

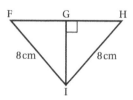

5 Lines AB and DE are parallel. Prove that triangles ABC and EDC are congruent.

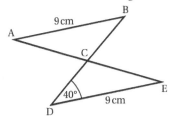

6 ABCD is a rectangle. E is the midpoint of AB. Prove that triangles AED and BEC are congruent.

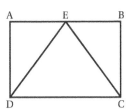

7 The shape ABCDE is a regular pentagon. The point F is the midpoint of CD.

Prove that the triangles ABG and AEG are congruent.

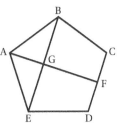

8 ABCD is a parallelogram.

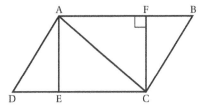

Find three pairs of congruent triangles. Justify your answer

Section 2: Applications of congruency
HOMEWORK 30B

1 In the diagram below, prove that triangle ABC is congruent to triangle ADC.

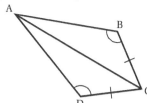

2 What shape is PQRS? Justify your answer.

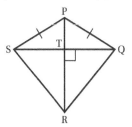

3 **a** In the diagram opposite, what type of triangle is ABD?
  b What is the size of the angle DCB?
  c What shape is ABCD?

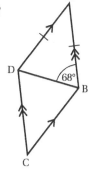

4 ABCDEF is a regular hexagon.
  a Which triangles are congruent?
  b What is the size of the angle FAD?
  c Which other angle is the same size as FAD?

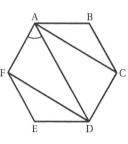

5 ABCD is a parallelogram. Prove that the angle ADC = ABC.

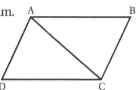

6 In the diagram below, AC = AB = BC. The point D is the midpoint of AB, E is the midpoint of AC and F is the midpoint of BC.
  a Which triangle is congruent to triangle ADC?
  b Write down four other triangles that are congruent.
  c Which triangle is congruent to triangle ECD?

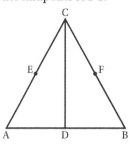

7

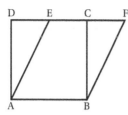

ABCD is a square. Point E is the midpoint of DC. BF is parallel to AE.
Prove that triangle ADE is congruent to triangle BCF.

8 Two parallelograms, ABCD and ABXY are on the same base. Prove that DCXY is a parallelogram.

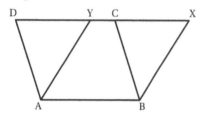

Chapter 30 review

1 Which of the following pairs of triangles are congruent? Give reasons for your answers. (All lengths are in centimetres.)

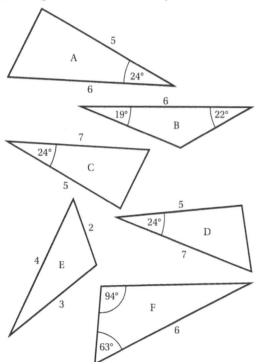

2 In the diagram below, TP = TQ. PR = QS. The angle TPQ = the angle TQS. Prove that the angle QRT = QST.

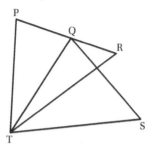

Tip

Prove that triangles TPR and TQS are congruent.

3 The triangle ABC is isosceles, with AB = AC. CE = DB.

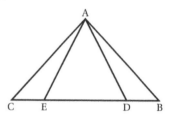

Prove that the angle ADE = AED.

4 The triangle ABC is equilateral. The sides AC and BC are extended and two new points drawn so that BD = 2BC and AE = 2AC. Join points D and E.
Draw a diagram to show this, and prove that the triangle BAD is right-angled.

5 ABCD is a rectangle. The point E is the midpoint of AC, F is the midpoint of BD, G is the midpoint of AB and H is the midpoint of CD.

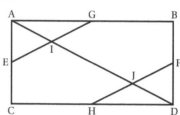

Prove that triangles HJD and GIA are congruent.

31 Similarity

Section 1: Similar triangles
HOMEWORK 31A

Tip

Similar means exactly the same shape, but a different size. All circles are similar.

① Each diagram opposite contains a pair of similar triangles. Use the correct terminology to identify the matching angles and the sides that are in proportion.

a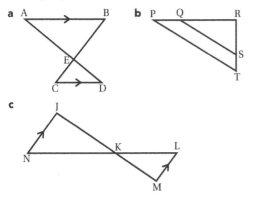

b

c

② Are the following pairs of triangles similar? Explain your reasoning.

a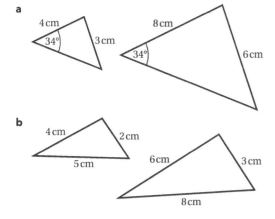

b

c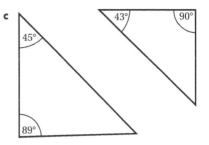

③ The two triangles below are similar. Find the missing lengths a and b.

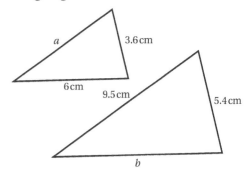

④ The two triangles below are similar. Find the missing lengths c and d.

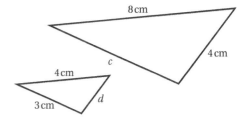

⑤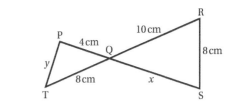

The triangles PQT and SQR are similar.
a Calculate the length of side x.
b Calculate the length of side y.

6 The side BC is parallel to the side DE.
Calculate the following lengths.

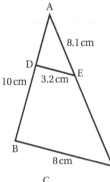

a AD
b DB
c AC

7

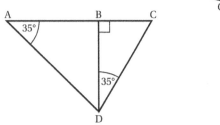

a State which pair of triangles in the diagram above is similar.
b Give reasons for your answer.

8 Two radio masts are set at right angles to the ground. Both cast a shadow at 1.00 p.m.
The longer mast is 80 m high and casts a shadow of 45 m.
The shorter mast is 55 m high. How long will its shadow be?
Give your answer to the nearest centimetre.

Section 2: Enlargements
HOMEWORK 31B

> **Tip**
>
> Remember to increase each side by the correct scale factor.

1 Enlarge each shape as directed.
a Enlarge by a scale factor of 2.

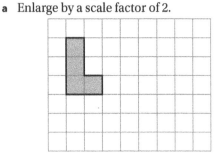

b Enlarge by a scale factor of 1.5.

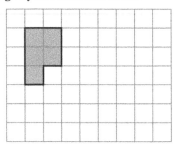

c Enlarge by a scale factor of $\frac{1}{2}$.

HOMEWORK 31C

1 Enlarge each shape using the scale factor given and the centre of enlargement shown.
a Scale factor 2.

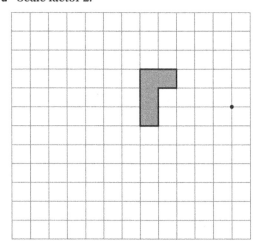

b Scale factor 1.5.

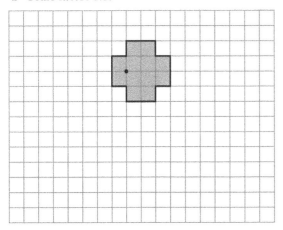

c Scale factor $\frac{1}{2}$.

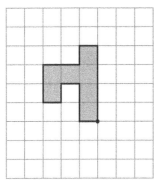

2 Enlarge the triangle by a scale factor of 1.5, using the origin as the centre of enlargement.

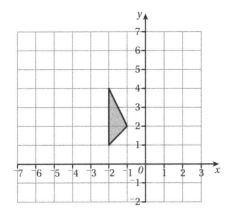

3 Enlarge the shape below by a scale factor of enlargement of $\frac{1}{2}$ with the centre of enlargement (⁻5, 3).

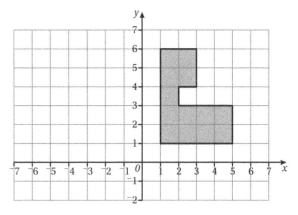

4 Enlarge the shape below by a scale factor of 2 with the centre of enlargement (0, 3).

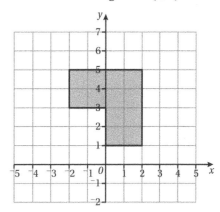

HOMEWORK 31D

1 Enlarge the given shape by a scale factor of ⁻1.5, using (1,0) as the centre of enlargement.

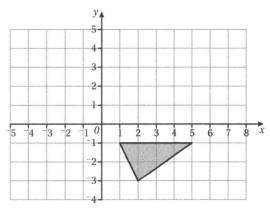

2 Enlarge the given shape by a scale factor of ⁻2, using (2, ⁻1) as the centre of enlargement.

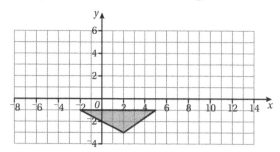

3 Enlarge the given shape by a scale factor of ⁻0.5, using (2, 2) as the centre of enlargement.

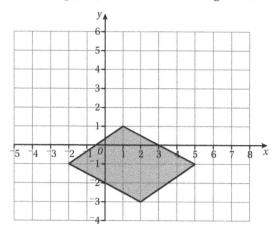

4 Enlarge the given shape by a scale factor of $\frac{-1}{2}$, using (⁻1,1) as the centre of enlargement.

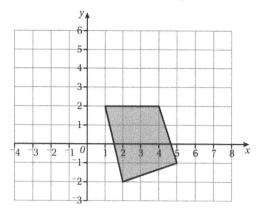

HOMEWORK 31E

1 Which of the following photographs is an enlargement of the original?

a   **b**

c

2 Describe each of the following enlargements, giving the scale factor and the centre of enlargement. The darker shapes are the original shapes and the lighter shapes are the enlargements.

a

b

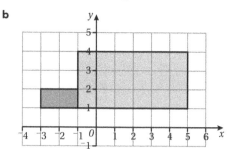

c

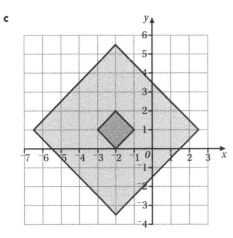

3 These diagrams each show an object and an image after enlargement. Describe each of these enlargements by giving both the scale factor and the coordinates of the centre of enlargement.

a

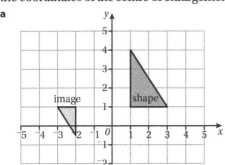

b

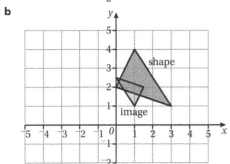

Section 3: Similar shapes and objects
HOMEWORK 31F

1 Decide whether each statement is true or false. Explain your reasoning.
   a All rectangles are similar.
   b All rectangles whose length is twice its width are similar.
   c All regular hexagons are similar.
   d All equilateral triangles are similar.
   e All isosceles triangles are similar.

2 Decide if the following pairs of shapes are similar. Explain your reasoning.
   a Rectangle ABCD with AB = 6 and BC = 4 cm.
      Rectangle EFGH with EF = 9 and FG = 6 cm.
   b Rectangle ABCD with AB = 10 and BC = 14 cm.
      Rectangle EFGH with EF = 7 and FG = 11 cm.
   c Rectangle ABCD with AB = 12 and BC = 9 cm.
      Rectangle EFGH with EF = 8 and FG = 6 cm.

3 The shapes below are similar. Find the lengths of a and b.

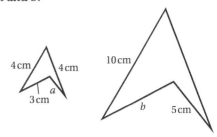

4 In the diagram below the shape PQRS has been created from enlarging ABCD by a scale factor of 2.5.

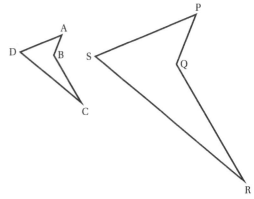

If AB = 2.5 cm, BC = 6 cm, CD = 7 cm and DA = 3.5 cm, find the lengths PQ, QR, RS and SP.

5 On a plan, a large square building has side lengths of 15 mm. In reality the building is a 750-metre square.
   a What is the scale factor of enlargement of the plan?
   b Write this scale as a ratio.

6 Two similar shapes have volumes in the ratio 1 : 216.
What is the ratio of their surface areas?

7 Two triangles A and B are similar. Triangle A has an area of 22.5 cm² and Triangle B has an area of 202.5 cm².
a What is the linear scale factor of enlargement?
b If the perpendicular height of Triangle A was 5 cm, what are the dimensions of Triangle B?

8 Baked beans are sold in large and small tins, which are similar cylinders.
A small tin has a radius of 3 cm and a height of 5 cm.
A large tin has a volume of 572 cm³.
Calculate the dimensions of the large tin.

9 A square-based pyramid has a base of side 5 m and a perpendicular height of 7 m.
a Calculate the volume of the pyramid.
b If a similar shape has a volume of 466.66 m³, what is the linear scale factor of enlargement between the smaller and larger objects?

Chapter 31 review

1 **a** Prove that the triangle ABC is similar to the triangle ADE.

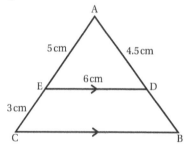

b Find the lengths AD, AB and BC.

2 A tent manufacturer makes similar shaped tents in three sizes.
A small tent has a width of 1.5 m and a height of 2 m.
A medium tent is 1.8 m wide, and a large tent is 2.7 m high.
a How high is a medium tent?
b How wide is a large tent?

3 Draw an enlargement of this shape, scale factor $\frac{1}{2}$ using the centre of enlargement shown.

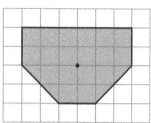

4 Draw an enlargement, scale factor 1.5 of this shape with the centre of enlargement at (0, 4)

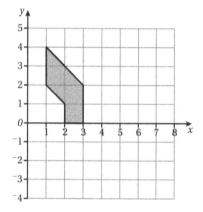

5 Are all parallelograms similar shapes? Explain.

6 Are all kites similar shapes? Explain.

7 A triangle has an area of 76.5 cm². A similar triangle has an area of 478.125 cm².
What is the linear scale factor of enlargement between the smaller and larger triangles?

8 A cuboid has a volume of 432 cm³. If a similar cuboid has a volume of 54 cm³, what is the linear scale factor of enlargement between the first and second cuboid?

32 Pythagoras' theorem

Section 1: Understanding Pythagoras' theorem

HOMEWORK 32A

1. Find the length of the hypotenuse in each of these right-angled triangles. Give your answers correct to three significant figures where appropriate.

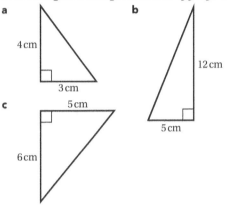

a   4 cm   3 cm

b   12 cm   5 cm

c   5 cm   6 cm

2. Find the length of the unmarked side in each of these right-angled triangles. Give your answers correct to three significant figures where appropriate.

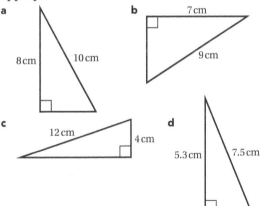

a   8 cm   10 cm

b   7 cm   9 cm

c   12 cm   4 cm

d   5.3 cm   7.5 cm

3. This diagram shows a kite. Use your knowledge of shapes and Pythagoras' theorem to find the missing lengths in the diagram.

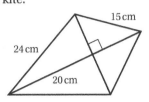

15 cm   24 cm   20 cm

4. The diagram below represents a perpendicular radio mast and two wires which are attached to the top of the mast.

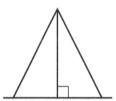

If each wire is 45 m long and the mast is 32.6 m high, how far apart are the two wires on the ground?

5. A new pyramid is discovered in Egypt, and explorers are able to measure how wide it is at the base and the length of the sloping side as shown in the diagram.

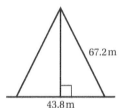

67.2 m   43.8 m

How high above the ground is the top of the pyramid?

6. A right-angled triangle has two sides of length 10 and 12 cm.
What are the two possible lengths of the other side in this triangle?

7. A sail on a sailing ship is a right-angled triangle. If the sail is 5.6 m wide at its base and has a hypotenuse of 11.6 m, how tall is it?

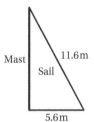

Mast   11.6 m   Sail   5.6 m

Section 2: Using Pythagoras' theorem

HOMEWORK 32B

1 Which of these are Pythagorean triples?
  a 5, 12, 13   **b** 12, 35, 37   **c** 12, 16, 20
  d 7, 25, 25   **e** 44, 117, 125

2 A builder is checking whether he has made a proper right-angle on some brickwork.
He measures two lengths of 4.5 m and a hypotenuse of 6.36 m.
Has he managed a correct right-angle?

3 A triangular sail has its hypotenuse of 8 m attached to the mast of a ship.
The other two sides of the sail are 3.4 m and 5.2 m long.
Is the sail right-angled?

4 Is a triangle with sides 8 cm, 15 cm, and 17 cm right-angled?

5 Which of these triangles are right-angled?
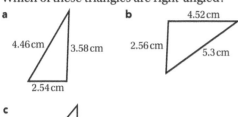

HOMEWORK 32C

> **Tip**
>
> Drawing a sketch often helps.

1 What is the perpendicular height of an isosceles triangle that has two equal sides of 12 cm and a third side of 15 cm?

2 The diagonal across a square is 14 cm.
What is the side length of the square?

3 A triangle has side lengths 12.2 cm, 13.5 cm and 18.5 cm. Is it a right-angled triangle?

4 A doorway is 1.9 m high and 1.2 m wide. Is it possible to fit a board 2.25 m wide through the door?

5 Farm gates are made of equally spaced parallel pieces of wood joined with a diagonal support. Given the dimensions in the diagram below, how long is the diagonal piece of wood?
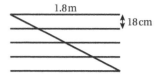

6 A window cleaner has a 5.3 m ladder. If he rests his ladder 5.1 m up a wall, how far away from the base of the wall will his ladder be (to the nearest cm)?

7 A radio mast is set perpendicular to the ground. The mast is 85 m tall. One of the cables securing the mast is 24 m away from the base of the mast.
  a How long is the cable?
  b A second cable is attached to the mast 65 m above the ground. The cable is 82 m long. How far away from the base of the mast is the cable?
Give your answers to the nearest cm.

8 The diagram below shows a square which is a cyclic quadrilateral.
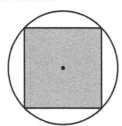
  a If the radius of the circle is 8 cm, what is the perimeter of the square?
  b What is the area of the square?

9 Find the value of the hypotenuse.
Leave your answer as a surd in simplified form.
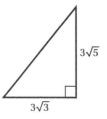

Section 3: Pythagoras in three dimensions
HOMEWORK 32D

1 A take-away coffee cup is a cylinder as shown in the diagram.
What is the length of the longest stirrer that could fit inside the cup if a flat lid is placed on top?

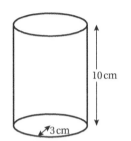

10 cm

3 cm

2 What is the shortest distance between two opposite vertices of a 5 m cube?

3 A light shade is a truncated cone as shown in the diagram. What is the perpendicular height of the light shade?

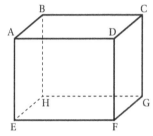

6 cm

15 cm

10 cm

4 For the cuboid shown, AB = 5 cm, AD = 7 cm and AE = 6 cm.

A new point I is formed at at the intersection of AG and BF.
How far is I from point D?

Section 4: Using Pythagoras to solve problems
HOMEWORK 32E

1 On a computer screen a character is 354 pixels to the right of the bottom-left corner and 213 pixels above it.
What is the shortest distance in pixels from the character to the bottom-left corner of the screen?

2 Photographic enlargements are sold in the following sizes:
a 6×4 inch **b** 7×5 inch **c** 8×10 inch
Work out the length of the diagonal for each size of photograph.

3 A TV has a screen size of 32 inches across the diagonal. The screen is 24 inches high, and has a border around it of 1 inch.
What is the narrowest gap this TV could fit into?

4 The front view of a building is shown in the diagram. How tall is the building?

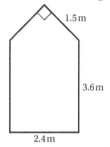

1.5 m

3.6 m

2.4 m

5 A football pitch measures 110 m by 60 m. What is the shortest distance across the pitch diagonally?

6 A surveyor uses a clinometer to measure the top of a tree to be 25.8 m away from him. The clinometer is 1.5 m above the ground and the base of the tree is 15.6 m away. How tall is the tree?

7 Will a 10 cm drinks stirrer fit inside a cylindrical container that is 7 cm high and has a radius of 1.6 cm?

8 A rower has an oar that is 6.2 m long. The back of his car, with the seats down, measures $3.5 \times 2.6 \times 5.1$ m.
Is there a way he can fit the oar in?

9 Two storage boxes have dimensions as follows:
Box A 6.4 cm $\times 4.3$ cm $\times 5.3$ cm
Box B 7.1 cm $\times 2.3$ cm $\times 4.8$ cm
Into which box could you fit the longest pole?

Chapter 32 review

1 Which of the following triangles are right-angled?
a 5 cm, 6 cm 8 cm
b 9 cm, 12 cm, 15 cm
c 4 cm, 4 cm, 8 cm

2 **a** What is the perimeter of this square if the length AB is 4.1 cm?

b What is the area of the square?

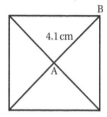

3 A ship sails 17.6 km due east then 15.4 km due south.

What is the shortest distance back to where it started?

4 A window cleaner wants to buy a new ladder. He estimates that the highest window that he needs to be able to reach is 8.4 m above the ground. The safety instructions state that the base of the ladder should be positioned 1 m away from the wall for every 4 m the ladder goes up vertically. What is the minimum length of ladder that he needs to buy? Give your answer in metres correct to one decimal place.

5 A light fitting is suspended from four chains attached in a square shape on the ceiling. The side length of the square is 12 cm. If the light needs to hang 18 cm below the ceiling, how long does each chain need to be?

6

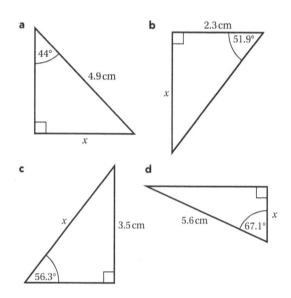

The diagram shows a cube. The length of the diagonal AB is 12 cm.

Work out the total surface area of the cube.

33 Trigonometry

Section 1: Trigonometry in right-angled triangles

HOMEWORK 33A

1 For the angle shown in this triangle, find:

a the sine ratio.

b the cosine ratio.

c the tangent ratio.

2 For each triangle, choose the appropriate ratio and find the length of the side indicated by the letter x. Give your answers correct to three significant figures.

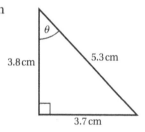

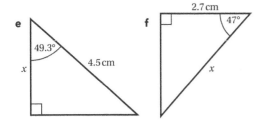

3 Find the length of the side marked x in each case.

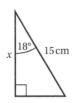
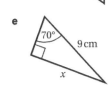
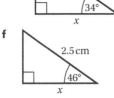

HOMEWORK 33B

1 Calculate the value of each angle given the following ratios:

a $\cos \theta = 0.457$ b $\tan \theta = 2.67$
c $\sin \theta = 0.867$ d $\tan \theta = 0.896$
e $\sin \theta = 0.014$ f $\cos \theta = 0.123$

2 For each triangle, choose the appropriate ratio and find the size of the angle indicated. Give your answers correct to one decimal place.

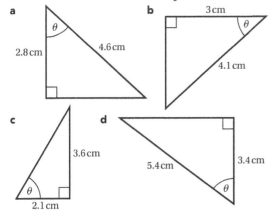

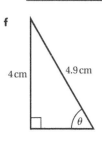
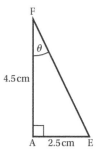

3 The triangle FEA is a right-angled triangle with
AE = 2.5 cm and
AF = 4.5 cm.
Find the size of the angle AFE.

4 What is the size of angle x?

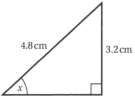

5 For each triangle, draw a sketch and then calculate the required values.
Give your answers correct to two decimal places.
a In triangle ABC, the angle A is 90°, the angle C = 40° and the side AB = 7.5 cm. Find the length of BC.
b In triangle PQR, the angle R is 90°, the angle P = 63° and the side PQ = 15.2 cm. Find the length of RQ.
c In triangle ABC, the angle B is 90°, the angle C = 31.4° and the side AB = 17.3 cm. Find the length of BC.

6 For each triangle, draw a sketch and then calculate the required values.
Give your answers correct to two decimal places.
a In triangle XYZ, the angle Y is 90°, YZ = 6.3 cm and XZ = 18.4 cm. Find the angle X.
b In triangle ABC, the angle A is 90°, AB = 8.9 cm and BC = 11.3 cm. Find the angle B.
c In triangle PQR, the angle R is 90°, PR = 17.3 cm and RQ = 18.4 cm. Find the angle P.

113

Section 2: Exact values of trigonometric ratios

HOMEWORK 33C

Tip

If you don't know these, you should learn them.

1 What is:
   a $\cos 30° + \sin 60°$?   **b** $\cos 60° + \sin 30°$?
   c $\cos 45° + \sin 45°$?

2 Write down the exact value of:
   a $\sin 0°$ **b** $\cos 30°$ **c** $\sin 60°$
   d $\cos 90°$ **e** $\sin 30°$ **f** $\tan 45°$
   g $\cos 0°$ **h** $\tan 60°$ **i** $\sin 45°$

3 Draw the two special cases of right-angled triangles containing the acute angles 30°, 45° and 60°. Use these to write down the exact values of:
   a $\tan 30°$ **b** $\cos 45°$
   c $\sin 60°$ **d** $\cos 60°$

4 Without using a calculator, determine the value of:
   a $\sin 45° \times \cos 45° + \sin 30°$
   b $(\tan 60°)^2$
   c $\sin 60° \times \cos 30° - \cos 60° \times \sin 30°$

Section 3: The sine, cosine and area rules

HOMEWORK 33D

1 Find the value of x in each of the following equations.
   Give your answers to two significant figures.

   a $\dfrac{x}{\sin 50°} = \dfrac{9}{\sin 38°}$

   b $\dfrac{x}{\sin 25°} = \dfrac{20}{\sin 100°}$

   c $\dfrac{20.6}{\sin 50°} = \dfrac{x}{\sin 70°}$

   d $\dfrac{\sin x}{11.4} = \dfrac{\sin 63°}{16.2}$

2 Find the length of the side marked x in each of the following triangles.
Give your answers to three significant figures.

a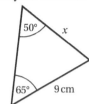

b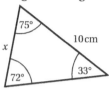

c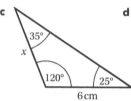

d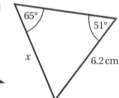

e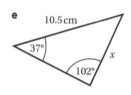

f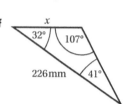

3 Find the size of the angle marked θ in each of the following triangles.
Give answers correct to one decimal place.

Tip

Some questions have two possible answers.

a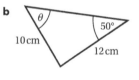

b

c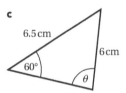

d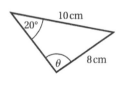

e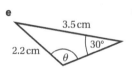

f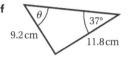

4 ABCD is a parallelogram with AB = 32 mm and AD = 40 mm. Angle BAC = 77°.

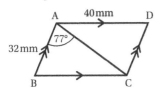

a Find the size of Angle BCA (to the nearest degree).
b Find the size of angle ABC (to the nearest degree).
c Find the length of diagonal AC correct to two decimal places.

HOMEWORK 33E

1 Find the length of the unknown side in each of these triangles. Give your answers to two significant figures.

a
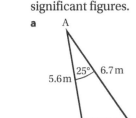

b
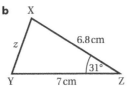

c
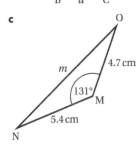

d
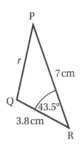

2 Find the size of angle *p* correct to one decimal place.

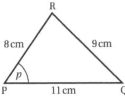

3 Given the following triangle, calculate:

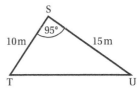

a the length of TU to three significant figures.
b the size of angle SUT to one decimal place.
c the size of angle STU to one decimal place.

4 Given that triangle ABC has AB = 6 cm, BC = 12 cm and AC = 15 cm, determine the size of angle ABC to the nearest degree.

5 Find the size of each angle in this triangle.

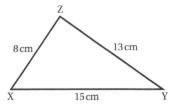

6 Given triangle DEF, find the size of angles E and F and the length DE.

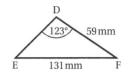

HOMEWORK 33F

1 Find the area of each triangle. Give your answers to three significant figures.

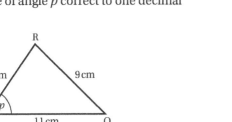

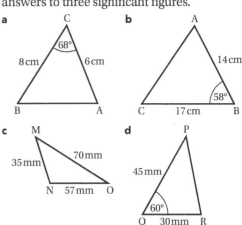

2 This triangle has an area of 20 cm². Find the size of angle F.

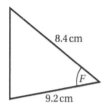

8.4 cm

F

9.2 cm

3 The diagram shows the plan of a small herb garden. Find the area of the garden correct to two decimal places.

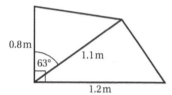

0.8 m

63°

1.1 m

1.2 m

4 Find the area of each polygon. Give your answers correct to one decimal place.

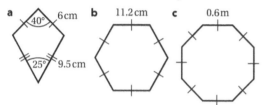

a 40° 6 cm **b** 11.2 cm **c** 0.6 m

25° 9.5 cm

Section 4: Using trigonometry to solve problems

HOMEWORK 33G

1 The coastguard is looking down from a cliff 78 m high through a telescope to a boat out at sea. The angle of depression is 26.3° and the telescope is 1.7 m high.
How far out to sea is the boat?

2 A radio phone mast is 67.3 m tall. It is held in place by wires attached to the top of the mast that make an angle of 78.4° with the ground.
How long are the wires?

3 A disabled access ramp is 10 m long and rises 50 cm.
What angle does the ramp make with the ground?

4 An aircraft is climbing at a consistent angle of 29.8° to the ground for a distance of 1.8 km through the air.
What is the equivalent distance along the ground?

5 A 15 m tall tower is viewed from two positions, A and B. A and B are on a straight line, but on opposite sides of the tower.
If the angle of elevation from A is 37° and from B is 24°, how far is it from A to B to the nearest metre?

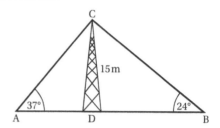

C

15 m

37°

24°

A   D   B

6 A ship sets sail from port on a bearing of 040°. After travelling 5 km in a straight line the ship makes a sudden turn and then travels 7 km on a bearing of 120°.
The ship then reaches point C.
Calculate:
a the direct distance between the point C and the port.
b the bearing on which the ship must set sail if it is to take the most direct route back to port.

7 Two coastguard stations are positioned 100 km apart on a piece of coastline that runs exactly east-west. A distress call is sent from a ship which reports that Station A is on a bearing of 330° from the ship and Station B is on a bearing of 080° from the ship.
Calculate:
a the bearings on which each of the two coastguards must set sail to reach the stricken ship.
b the distance that each of the two lifeboats will need to travel to provide help.
Give your answers to one decimal place.

8 A cuboid is 4 cm wide, 5 cm high and 8 cm long. Find the length of the longest diagonal of the cuboid.

9 A charity builds small homes for poor communities. The dimensions of a home are given in the diagram.
The vertex of the roof (V) is directly above the centre of the rectangular floor.

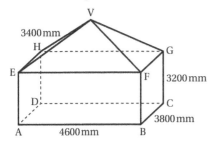

Find:
a the height of the house, in metres, from the centre of the floor to point V.
b the angle of elevation from point A to V.
c the distance from A to C.
d the distance from A to G.

Section 5: Graphs of trigonometric functions

HOMEWORK 33H

1 Two graphs, A and B, are drawn on the same system of axes.

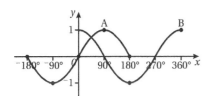

a Describe each graph by giving its equation and the values between which x lies.
b For which values is:
  i $\sin x = \cos x$
  ii $\sin x > \cos x$
c What is the value of $\cos x$ when $x = 0°$?
d Between which two values does y range for both graphs?
e Find the solution of the equation $\sin x = {}^-1$ from the graph.

2 Explain how the graph of $\tan x$ for values of x between $^-180°$ and $360°$ would differ from the graphs shown in question 1.

Chapter review 33

1 The diagram shows the triangle ABC. Find the size of the marked angle.

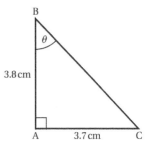

2 The triangle XYZ is shown in the diagram below.

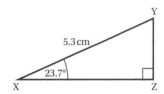

Find the length of the other two sides.

3 The diagram shows a riverbank.

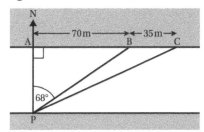

Ali is standing at point P and Alesha is standing due north of Ali, at point A on the other side of the river. Alesha walks to point B. She is now on a bearing of 068° from point P.
a Calculate the width of the river AP.
Give your answer correct to three significant figures.

Alesha now walks to point C.
b Calculate the bearing of point C from P.

4 Triangle ABC has AB = 7.2 cm and BC = 6.5 cm. Angle ABC is 140°.
Find the area of the triangle correct to two decimal places.

5 Find the length of AB in each triangle. Give your answers correct to three significant figures.

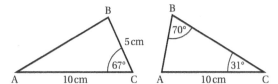

6 What is the size of angle XYZ in this triangle?

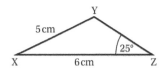

7 A 9 m high vertical mobile phone mast on a level concrete slab is supported by two stay wires 10 m long. Each stay wire is attached to the top of the pole and to the slab. Calculate:

  a the angle between the stay wire and the slab.
  b the distance from the bottom of the mast to the point where the stay wires are attached to the slab.

8 Calculate the angle of elevation of the Sun (to 1 dp) if a vertical wall 1 m high casts a 0.83 m shadow. Assume the ground is level.

9 In triangle ABC, BC = 9.8 cm, angle ABC = 32° and angle ACB = 75°. Find the lengths of AB and AC.

10 The diagram shows a triangular prism. The rectangular base ABCD is horizontal.

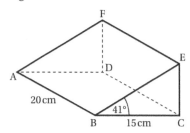

AB = 20 cm and BC = 15 cm
The cross-section of the prism BCE is right-angled at C and angle EBC is 41°.
  a Determine the length of AC.
  b How long is EC?
  c What angle does line AE make with the horizontal?

34 Circle theorems

Section 1: Review of parts of a circle
HOMEWORK 34A

1 Draw a circle with a radius of 4 cm and centre O. By drawing the parts and labelling them, indicate the following on your diagram:
  a a sector with an angle of 50°
  b chord DE
  c MON, the diameter of the circle
  d a tangent that touches the circle at M
  e the major arc MP.

2 Use the diagram of the circle with centre O to answer these questions.

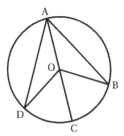

  a What are the correct mathematical names for:
    i DO?   **ii** AB?   **iii** AC?
  b Four radii are shown on the diagram. Name them.
  c If OB is 12.4 cm long, how long is AC?
  d Draw a copy of the circle and draw the tangent to the circle that passes through point B.

Section 2: Circle theorems and proofs
HOMEWORK 34B

Tip

You might need to use the angle relationships for triangles, quadrilaterals and parallel lines, as well as Pythagoras' theorem, to solve circle problems.

1. Given that O is the centre of the circle, calculate the value of angle BAD with reasons.

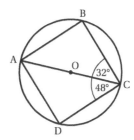

2. In the diagram, O is the centre of the circle, angle NPO = 15° and angle MOP = 70°.

Find the size of the following angles, giving reasons.
a. PNO
b. PON
c. MPN
d. PMN

3. O is the centre of the circle. Calculate the size of the following angles.
a. ACD
b. AOD
c. BDC

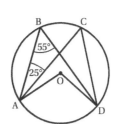

4. AB is the diameter of the circle. Calculate the size of angle ABD.

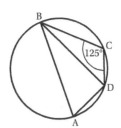

5. Find the size of angle x in each of the following diagrams.
Show your working and give reasons for any statements you make.

a
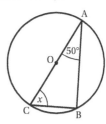

b
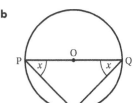

c
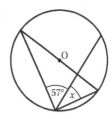

HOMEWORK 34C

1. Determine the length of the chord shown in each diagram.

a
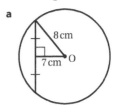

b
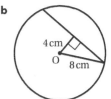

c
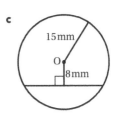

2. Find the size of the angles marked x and y.

a
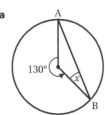

b
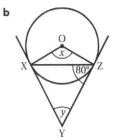

3 In the diagram, MN and PQ are equal chords. S is the midpoint of MN and R is the midpoint of PQ. MN = 12.5 cm. Find the length of SO correct to two significant figures.

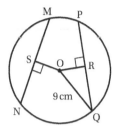

4 a MN and PQ are equal chords and MN = 12 cm. S and T are the midpoints of MN and PQ respectively. SO = 5 cm.

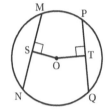

b Find the length of the diameter and then determine its circumference correct to two decimal places. Chord AB = 46 mm

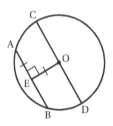

5 Find the size of angle *x* in each diagram. Give reasons for any statements you make.

a

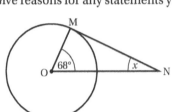

b

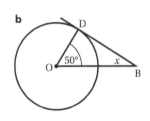

c

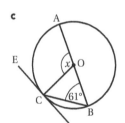

d

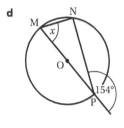

e

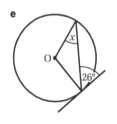

HOMEWORK 34D

1 Find the size of angle *x* in each circle. Give reasons for any statements you make.

a

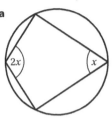

b

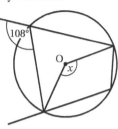

2 Calculate the angles of the cyclic quadrilateral.

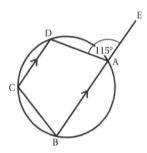

3 Given that O is the centre of the circle and angle BAC = 72°, calculate the size of angle BOC.

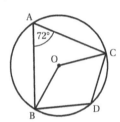

4 In the diagram, AB//DC, angle ADC = 64° and angle DCA = 22°.

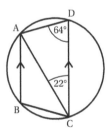

Calculate the size of:
a angle BAC
b angle ABC
c angle ACB.

Section 3: Applications of circle theorems

HOMEWORK 34E

1. SNT is a tangent to a circle with centre O. Angle QMO = 18° and angle MPN = 56°.

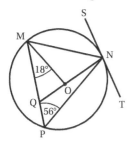

Find the size of:
a angle MNS
b angle MOQ
c angle PNT.

2. Find the size of angles *a* to *e*.

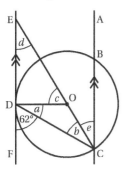

3. The diagram shows a circular tabletop. A spider walks from point A to point B, and the closest that the spider gets to the centre of the circle is 18 cm. The spider then walks directly from the point B to C, once again taking a route such that the closest the spider comes to the centre of the circle is 18 cm.

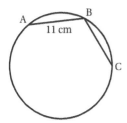

Calculate the total distance walked by the spider.

4. The diagram shows a circle with two tangents PX and XQ. The tangents intersect at the point X and meet the circle at P and Q respectively.

If the centre of the circle is O, and the angle POQ is 150°, calculate the angle QPX.
Give clear reasons for each step of your working.

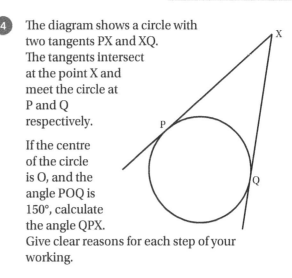

Chapter 34 review

1. In each of the following O is the centre of the circle. Find the value of the marked angles. Give reasons for your statements.

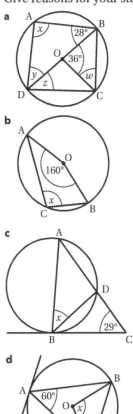

121

2 Find the length of x and y in each of these diagrams.
O is the centre of the circle in each case.

a

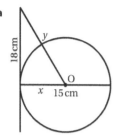

b

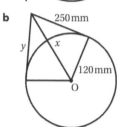

3 SPT is a tangent to a circle with centre O.
SR is a straight line which goes through the centre of the circle and angle PSO = 29°.
Find the size of angle TPR.

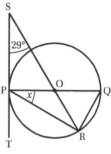

4 Work out the size of angle x. Give reasons for your answer.

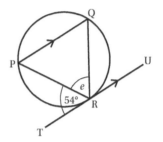

35 Discrete growth and decay

Section 1: Simple and compound growth

HOMEWORK 35A

1 Calculate the simple interest on:
a £250 invested for a year at the rate of 3% per annum.
b £400 invested for five years at the rate of 8% per annum.
c £700 invested for two years at the rate of 15% per annum.

2 £7500 is invested at 3.5% per annum simple interest.
How long will it take for the amount to reach £8812.50?

3 The total simple interest on £1600 invested for five years is £224.
What is the percentage rate per annum?

4 Calculate the compound interest on:
a £250 invested for a year at the rate of 3% per annum.
b £400 invested for five years at the rate of 8% per annum.
c £700 invested for two years at the rate of 15% per annum.

5 How much will you have in the bank if you invest £500 for four years at 3% interest, compounded annually?

6 Mrs Genaro owns a small business. She borrows £18 500 from the bank to finance some new equipment. She repays the loan in full after two years. If the bank charged her compound interest at the rate of 21% per annum, how much did she repay over the two years?

7 Rupert's grandmother gives him £x on his 18th birthday. If Rupert leaves the money in an account that gains 4% interest per year, and if he does not make any withdrawals, Rupert's grandmother will add another £x on his next birthday. This will keep happening for every year that the money is left in the account.

a If Rupert leaves the money untouched, how much money will be in his account one day after his 19th birthday?

b How will much there be after his 20th birthday?

c If Rupert finds that there is £849 292.80 in his account one day after his 21st birthday, find the value of x.

Section 2: Simple and compound decay
HOMEWORK 35B

1 A computer system cost £9500 to install. The system decreases in value by £1500 per annum. What is its value after four years?

2 A car costing £10 000 depreciates in value by 10% per year.
What is the car worth on paper after seven years?

3 The number of applications at a college is reducing by 8% per annum. If 3800 applications were received in 2014, how many applications would you expect to have in 2018?

4 In 2010 there were an estimated 1600 giant pandas in China. Calculate the likely panda population in 2025 if there is:
a an annual growth in the population of 0.5%.
b an annual decline in the population of 0.5%.

5 A security system costs £8400 to install. If the company calculates depreciation on a reducing balance at 15% per annum, what is the system worth four years after it is installed?

6 A piece of equipment has a value of £13 770 after 10 years. If it depreciated at a rate of 8.2% per annum, what was the original cost of the equipment?

Chapter 35 review

1 A woman invests £5000 in an investment scheme for five years and earns 8% pa simple interest.
a Calculate the total interest she will earn.
b How much would she need to invest to earn £3600 interest in the same period (at the same rate)?
c At the end of an investment period, the woman is told her £5000 has increased in value by 23%. Use a multiplier to work out how much the investment is worth at the end of the period.

2 This table compares the simple and compound interest earned on £10 000 invested at a rate of 9% pa.

Years	Simple interest	Compound interest
1	900	900
2	1800	1881
3	2700	2950.29
4	3600	4115.82
5	4500	5386.24
6	5400	6771.00
7		
8		

a Copy and complete the last two rows of the table.
b What is the difference between the simple interest and compound interest earned after five years?

3 An investment of £9500 does very badly and the client loses money at a rate of 20% per annum. What is the investment worth after four years?

4 Michelle buys a new car for £39 000. Her dad tells here it's a poor investment. He claims it loses 23% of its value as soon as she drives it out of the showroom and thereafter it is worth 10% less each year she owns it.
If her dad is right, how much value has Michelle's new car lost at the end of two years?

5 A computer system depreciates at a rate of 20% per annum. Its purchase price was £8500.
After how many years will it be worth approximately £2000?

36 Direct and inverse proportion

Section 1: Direct proportion

HOMEWORK 36A

1   Determine whether A and B are directly proportional in each case.

a

A	2	4	6
B	300	600	900

b

A	1	5	8
B	2	10	15

c

A	1	2	3	4
B	0.1	0.2	0.3	0.4

2   Find the cost of five identically priced items if seven items cost £17.50.

3   If a 3.5 m tall pole casts a 10.5 m shadow, find the length of the shadow cast by a 20 m tall pole at the same time.

4   A truck uses 20 litres of diesel to travel 240 kilometres.
   a How much diesel will it use to travel 180 km at the same rate?
   b How far could the truck travel on 45 litres of diesel at the same rate?

Section 2: Algebraic and graphical representations

HOMEWORK 36B

1   This graph shows the directly proportional relationship between lengths in metres (metric) and lengths in feet (imperial).

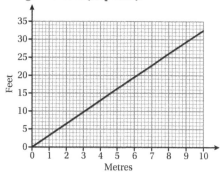

a Use the graph to estimate how many feet there are in 4 metres.
b Given that 1 m = 3.28 feet, calculate how many feet there are in 4 metres.
c Which is longer:
   i 4 metres or 12 feet?
   ii 20 feet or 6.5 metres?
d Mr Bokomo has a length of fabric 9 m long.
   i What is its length to the nearest foot?
   ii He cuts and sells 1.5 m to Mrs Johannes and 3 ft to Mr Moosa. How much is left in metres?
e A driveway was 18 feet long. It was resurfaced and extended to be one metre longer than previously. How long is the newly resurfaced driveway in metres?

Tip

When two quantities are directly proportional then $P = kQ$, where k is a constant.

2   Given that a varies directly with b and that $a = 56$ when $b = 8$,
   a find the value of the constant of proportionality (k),
   b find the value of a when $b = 12$.

3   The table shows values of m and T. Show that T is directly proportional to m.

m	7.5	11.5	18
T	16.35	25.07	39.24

4   F is directly proportional to m and $F = 16$ when $m = 2$.
   a Find the value of F when $m = 5$.
   b Find the value of m when $F = 36$.

5   $y = kx$. When $y = 24$, $x = 16$. Calculate:
   a the value of k.
   b y when $x = 10$.
   c x when $y = 12$.

Section 3: Directly proportional to the square, square root and other expressions

HOMEWORK 36C

1 y is directly proportional to x^2, and $y = 50$ when $x = 5$.
  a Write the equation for this relationship.
  b Find y if $x = 25$.
  c Find x if $y = 162$.

2 y is directly proportional to x^3 and when $x = 2$, $y = 32$.
  a Write this relationship as an equation.
  b Find the value of y when $x = 5$.

3 $y \alpha \sqrt{x}$. When $y = 25$, $x = 25$. Find:
  a y when $x = 16$.   **b** x when $y = 2.5$.

4 The speed of water in a river is determined by a water-pressure gauge.
The speed (v m/s) is directly proportional to the square root of the height (h cm) reached by the liquid in the gauge. Given that $h = 36$ when $v = 8$, calculate the value of v when $h = 18$.

5 You are told that y is directly proportional to the square of x, x is directly proportional to the cube of z and $y = 128$ when $z = 2$.
Find the value of y when $z = 4$.

Section 4: Inverse proportion

HOMEWORK 36D

1 It takes one employee ten days to complete a project. If another employee joins him, it only takes five days. Five employees can complete the job in two days.
  a Describe this relationship.
  b How long would it take to complete the project with:
    i four employees?   **ii** 20 employees?

2 A plane travelling at an average speed of 1000 km/h takes 12 hours to complete a journey.
How fast would it need to travel to cover the same distance in ten hours?

3 A journey takes three hours when you travel at 60 km/h. How long would the same journey take at a speed of 50 km/h?

4 For each of the following, y is inversely proportional to x.
Write an equation expressing y in terms of x if
  a $y = 0.225$ when $x = 20$
  b $y = 12.5$ when $x = 5$ **c** $y = 5$ when $x = 0.4$
  d $y = 0.4$ when $x = 0.7$ **e** $y = 0.6$ when $x = 8$.

5 F is inversely proportional to d^2 and when $d = 3$, $F = 12$.
Find the value of F when $d = 4$.

Chapter 36 review

1 A car used 45 litres of fuel to travel 495 km.
  a How far could the car travel on 50 l of fuel at the same rate?
  b How much fuel would the car use to travel 190 km at the same rate?

2 It takes six people 12 days to paint a building. Work out how long it would take at the same rate using:
  a 9 people.   **b** 36 people.

3 Study the graph and answer the questions.

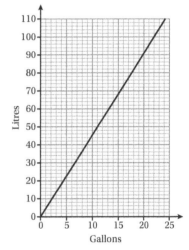

  a What does the graph show?
  b Convert to litres.
    i 10 gallons   **ii** 25 gallons
  c Convert to gallons.
    i 15 litres   **ii** 120 litres
  d Naresh says he gets 30 mpg in the city and 42 mpg on the motorway in his car.
    i Convert each rate to km per gallon.
    ii Given that one gallon imperial is equivalent to 4.546 litres, convert both rates to kilometres per litre.

4 A is directly proportional to r^2 and when $r = 3$, $A = 36$. Find the value of A when $r = 10$.

5 I is inversely proportional to d^3. When $d = 2$, $I = 100$. Find the value of I when $d = 5$.

6 An electric current I flows through a resistance R. I is inversely proportional to R and when $R = 3$, $I = 5$. Find the value of I when $R = 0.25$.

7 When an object falls freely under gravity, the force on the object due to air resistance F (measured in Newtons, N) is directly proportional to the square of the velocity v (measured in metres per second). When the object falls at 10 metres per second, the air resistance is 2.125N.

a Find a formula connecting velocity and air resistance force.

b An object will fall at terminal velocity when the weight of the object (the force due to gravity also measured in Newtons) is exactly balanced by the air resistance.
If an object has weight 850N, calculate the terminal velocity of the object when falling under gravity.

37 Collecting and displaying data

Section 1: Populations and samples
HOMEWORK 37A

1 A local shop-owner wants to find out how many boxes of a new crisp flavour she should order. She asks the first ten customers who come into the shop whether they would buy the new flavour if she started selling them.
   a What is the population in this survey?
   b What is the sample involved in the survey?
   c Is this a representative sample or not? Give a reason for your answer.

2 Sami says: 'More and more people are texting these days instead of phoning and talking to each other.' How could you collect data to find out whether this statement is true or not? Include details about the sources of your data and the sample size.

3 The statements below show you what four students found out when they collected data.

   Student A: 30% of heart disease is caused by smoking.

   Student B: 79% of all garbage is non-recyclable and is made up mostly of food and garden waste.

   Student C: Most people spend between £5 and £10 per day on transport.

   Student D: Almost $\frac{2}{3}$ of the women at my mum's workplace say that men earn more than they do.

   a What question do you think each student was trying to answer?
   b What sources of information do you think each student used to find their data?
   c How do you think student D selected a random sample from her mum's workplace?

Section 2: Tables and graphs
HOMEWORK 37B

1 Nika tossed a dice 40 times and got these results.

6	6	6	5	4	3	2	6	5	4
1	1	3	2	5	4	3	3	3	2
1	6	5	5	4	4	3	2	5	4
6	3	2	4	2	1	2	2	1	5

   a Copy and complete this frequency table to organise the data.

Score	Frequency
1	
2	
3	
4	
5	
6	

   b Do the results suggest that this is a fair dice or not? Give a reason for your answer.

2 Study the diagram carefully and answer the questions about it.

Number of students in each year

Year 8	𝕏 𝕏 𝕏 𝕏 𝕏 𝕏 𝕏 𝕏
Year 9	𝕏 𝕏 𝕏 𝕏 𝕏 𝕏
Year 10	𝕏 𝕏 𝕏 𝕏 𝕏 𝕏 𝕏
Year 11	𝕏 𝕏 𝕏 𝕏 𝕏 𝕏 𝕏 𝕏 𝕏
Year 12	𝕏 𝕏 𝕏 𝕏 𝕏 𝕏 𝕏 𝕏

Key:
𝕏 = 30 students

a What type of chart is this?
b What does the chart show?
c What does each full symbol represent?
d How are 15 students shown on the chart?
e How many students are there in year 8?
f Which year group has the most students? How many are there in this year group?
g Do you think these are accurate or rounded figures? Why?

3 The table shows the population (in millions) of five of the world's largest cities.

City	Population (millions)
Tokyo	32.5
Seoul	20.6
Mexico City	20.5
New York	19.75
Mumbai	19.2

Draw a pictogram to show this data.

4 The frequency table shows the number of people who were treated for road accident injuries in the A & E ward of a large hospital in the first six months of the year.
Draw a bar chart to represent the data.
Use a scale of 1 cm per **50** patients on the vertical axis.

Patients admitted as a result of road accidents	
Month	**Number of patients**
January	360
February	275
March	190
April	375
May	200
June	210

5 Draw a bar chart to show this data.

Favourite takeaway food	No. of people
Burgers	40
Kebabs	30
Fried chicken	84
Hot chips	20
Other	29

HOMEWORK 37C

1 The table below shows the type of food that a group of students on a camping trip chose for breakfast.

	Cereal	Hot porridge	Bread
Girls	8	16	12
Boys	2	12	10

a Draw a single bar chart to show the choice of cereal against bread.
b Draw a compound bar chart to show the breakfast food choice for girls and boys.

2 The favourite subjects of a group of students are shown in the table.

Subject	Girls	Boys
Mathematics	34	33
English	45	40
Biology	29	31
ICT	40	48

a Draw a double bar graph to show this data.
b How many girls chose mathematics?
c How many boys chose ICT?
d Which is the favourite subject among the girls?
e Which is the least favourite subject amongst the boys?

3 A tourist organisation in the Caribbean records how many tourists from the UK and other international regions visit the Caribbean each month. Draw a compound bar graph to display the data for the first six months of the year.

	UK visitors	Other international visitors
Jan.	12 000	40 000
Feb.	10 000	39 000
Mar.	19 000	15 000
Apr.	16 000	12 000
May	21 000	19 000
Jun.	2000	25 000

Section 3: Pie charts
HOMEWORK 37D

1. This pie chart shows the colours that 80 students selected as their favourite from a five-colour chart.

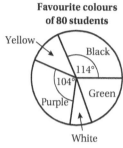

Favourite colours of 80 students

a Which colour is most popular?
b Which colour is least popular?
c What percentage of the students chose purple as their favourite colour?
d How many students chose black as their favourite colour?

2. This table shows the approximate percentage of the world's population living on each continent.

Africa	13
Asia	61
Europe	12
North America	5
South America	8.5
Oceania	0.5

a Draw a pie chart to display this data.
b How else could you display this data?

Section 4: Cumulative frequency curves and histograms
HOMEWORK 37E

1. This partially completed histogram shows the heights of trees in a section of tropical forest.

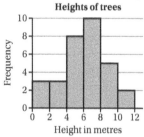

Heights of trees

a A scientist measured five more trees and their heights were
2.09 m, 3.34 m, 6.45 m, 9.26 m and 3.88 m.
Redraw the graph to include this data.
b How many trees in this sector of forest are ≥ 6 m tall?
c What is the modal class of tree heights?

2. A nurse measured the masses of a sample of students in a high school and drew the following table.

Mass kg	Frequency
$54 \leqslant m < 56$	4
$56 \leqslant m < 58$	7
$58 \leqslant m < 60$	13
$60 \leqslant m < 62$	19
$62 \leqslant m < 64$	11

a Draw a histogram to show the distribution of masses.
b What is the most common mass?
c What percentage of students weighed less than 56 kg?
d What is the maximum possible range of the masses?

3. The table shows the average minutes of airtime that teenagers bought from a pre-paid kiosk in one week.

Minutes	No. of teenagers
$20 \leqslant m < 30$	10
$30 \leqslant m < 40$	15
$40 \leqslant m < 60$	40
$60 \leqslant m < 80$	50
$80 \leqslant m < 100$	60
$100 \leqslant m < 150$	50

Draw an accurate histogram to display this data. Use a scale of 1 cm to represent ten minutes on the horizontal axis and an area scale of 1 cm² per five persons.

4. Thirty seedlings were planted for a biology experiment. The heights of the plants were measured after three weeks and recorded as below.

Heights h cm	Frequency
$0 \leqslant h < 3$	3
$3 \leqslant h < 6$	8
$6 \leqslant h < 9$	15
$9 \leqslant h < 12$	4

a Find an estimate for the mean height.
b Draw a cumulative frequency curve and use it to find the median height.
c Estimate Q1 and Q3 and the IQR.

Section 5: Line graphs for time series data

HOMEWORK 37F

1 Amy bought a new car in 2010. Its value over time is shown below.

Year	Value of car
2010	£13 900
2011	£7000
2012	£5700
2013	£4700
2014	£4000

  a Draw a line graph to represent this information.
  b What is the percentage depreciation in the first year she owned the car?
  c Use your graph to estimate the value of the car in 2015.

2 The table shows the distance (metres) covered by a car travelling at 90 km/h.

Time (in seconds)	Distance covered (in metres)
40	1000
80	2000
120	3000
160	4000
200	5000

Draw a line graph to show this relationship.

Chapter 37 review

1 Michelle collected data about how many children different families in her community had. These are her results.

```
0 3 4 3 3 2 2 2 2 1 1 1
3 3 4 3 6 2 2 2 0 0 2 1
5 4 3 2 4 3 3 3 2 1 1 0
3 1 1 1 1 0 0 0 2 4 5 3
```

  a How do you think Michelle collected the data?
  b Draw up a frequency table, with tallies, to organise the data.
  c Represent the data on a pie chart.
  d Draw a bar chart to compare the number of families that have three or fewer children with those that have four or more children.

2 Sally did a survey to find the ages of people using an internet cafe. These are her results:

Age (a)	No. of people
$15 \leqslant a < 20$	14
$20 \leqslant a < 25$	12
$25 \leqslant a < 35$	12
$35 \leqslant a < 50$	12
$50 \leqslant a < 55$	8

Draw an accurate histogram to show these data. Use a scale of 1 cm to five years on the horizontal axis and an area scale of one square centimetre to represent two people.

3 This histogram shows the number of houses in different price ranges that are advertised in a property magazine.

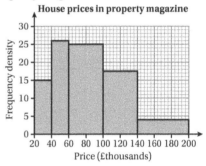

House prices in property magazine

  a How many houses were in the £20 000–£40 000 price range?
  b How many houses were in the £140 000–£200 000 price range?
  c How many houses are represented by one square centimetre on this graph?

4 This cumulative frequency curve shows the height in cm of 200 professional basketball players.

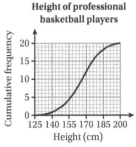

Height of professional basketball players

  a Estimate the median height of players in this sample.
  b Estimate Q1 and Q3.   **c** Estimate the IQR.
  d What percentage of basketball players are over 1.82 m tall?

38 Analysing data

Section 1: Summary statistics
HOMEWORK 38A

1 For each of the following frequency distributions calculate:
   a the mean score.
   b the median score.
   c the modal score.

Data set A

Score	1	2	3	4	5	6
Frequency	12	14	15	12	15	12

Data set B

Score	10	20	30	40	50	60	70	80
Frequency	13	25	22	31	16	23	27	19

Data set C

Score	1.5	2.5	3.5	4.5	5.5	6.5
Frequency	15	12	15	12	10	21

2 The table shows the number of words per minutes typed by a group of computer programmers.

Words per minute (w)	Frequency
$31 \leqslant w < 36$	40
$36 \leqslant w < 41$	70
$41 \leqslant w < 46$	80
$46 \leqslant w < 51$	90
$51 \leqslant w < 55$	60
$55 \leqslant w < 60$	20

   a Determine an estimate for the mean number of words typed per minute.
   b What is the modal number of words typed per minute?
   c What is the median class?
   d What is the range of words typed per minute?

3 For each of the following sets of data calculate the median, upper and lower quartiles. In each case calculate the interquartile range.
   a 67 44 63 56 46 48 55 63
   b 17 18 17 14 8 3 15 18 3 15
   c 0.8 1.3 0.7 1.4 2.3 0.4
   d 1 0 2 2 0 4 1 3 1 2 3 4 5 4 5 5

4 20 pupils take a French test and their scores out of 100 are recorded below.

   34 76 92 89
   21 23 45 87
   65 96 23 38
   72 91 32 77
   98 80 81 20

   a Find the median score for the pupils.
   b Calculate the interquartile range for the pupils' scores.

HOMEWORK 38B

1 Use the cumulative frequency curve below to find the missing values in each sentence.

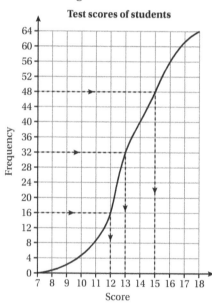

Test scores of students

   a __ students scored below 15. __ marks is the upper quartile.
   b 32 students scored below __. __ marks is the median or Q_2 mark.
   c 16 students scored below __. __ marks is the first quartile or Q_1.
   d The IQR is __.

2 The percentage scored by 1000 students on an exam is shown on this cumulative frequency curve.

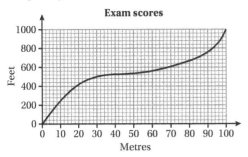

Exam scores

Use the curve to find an estimate for:
a the median score
b the lower quartile
c the upper quartile
d the interquartile range.

3 The lengths of 32 metal rods were measured and recorded on this cumulative frequency curve.

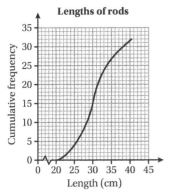

Lengths of rods

Use the graph to find an estimate for:
a the median   **b** Q_1
c Q_3 **d** the IQR.

HOMEWORK 38C

1 Five students scored a mean mark of 14.8 out of 20 for a maths test.
a Which of these sets of marks fit this average?
   i 14, 16, 17, 15, 17   **ii** 12, 13, 12, 19, 19
  iii 12, 19, 12, 18, 13   **iv** 13, 17, 15, 16, 17
   v 19, 19, 12, 0, 19   **vi** 15, 15, 15, 15, 14
b Compare the sets of numbers in your answer above.
Explain why you can get the same mean from different sets of numbers.

2 Twenty students scored the following results in a test out of 20.

17 18 17 14 8 3 15 18 3 15
0 17 16 17 14 7 18 19 5 15

a Calculate the mean, median, mode and range of the marks.
b Why is the median the best summary statistic for this particular set of data?

3 The table shows the times in minutes and seconds that two runners achieved over 800 m during one season.

Runner A	2 m 2.5 s	2 m 1.7 s	2 m 2.2 s	2 m 3.7 s	2 m 1.7 s	2 m 2.9 s	2 m 2.6 s
Runner B	2 m 2.4 s	2 m 1.8 s	2 m 2.3 s	2 m 4.4 s	2 m 0.6 s	2 m 2.2 s	2 m 1.2 s

a Which runner is the better of the two? Why?
b Which runner is most consistent? Why?

4 Two students get the following results for six mathematics tests out of 100.

Anna: 60, 90, 100, 90, 90, 100
Zane: 60, 70, 60, 70, 70, 100

a What is the range of scores for each student?
b Does this mean they both had equally good results?
c Which statistic would be a better measure of their achievement? Why?

5 For each of the box plots below identify the:
i median   **ii** range   **iii** upper quartile
iv lower quartile   **v** IQR.

A
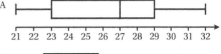

B
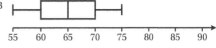

C
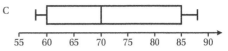

D
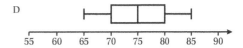

E
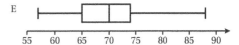

6 The weekly growth of 13 plants is given to the nearest millimetre.

12   13   5   8   9   24   16
14   14   6   9   12   12

a Draw a box plot to represent this data.
b What is the IQR of this data set?

7 The box plots below show the test results (out of 30) for two year 10 classes.

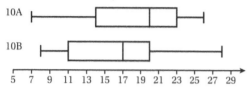

a Summarise the data for each class.
b What do these results tell you about the performance of the two classes?

Section 2: Misleading graphs
HOMEWORK 38D

1 This graph shows the number of computers sold by two competing stores over a four-month period.

a Did shop B sell double the number of computers that shop A sold in January? Give a reason for your answer.
b Did shop A sell four times as many computers as shop B in March? Explain.
c Calculate the total number of computers sold over the period for each shop. How do the figures compare?
d How is this graph misleading?

2 Study the pie graph.

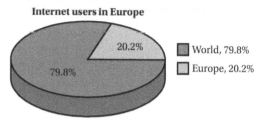

a What does the graph suggest to you? Why?
b Europe has 11.5% of the world's population. Does this affect how you interpret this graph? Explain.
c Why should pie graphs not be shown with 3D sections?

Section 3: Scatter diagrams
HOMEWORK 38E

1 a Describe the correlation shown on the following scatter diagrams.
b Draw a line of best fit on graphs **i**, **ii**, **iv** and **v**.

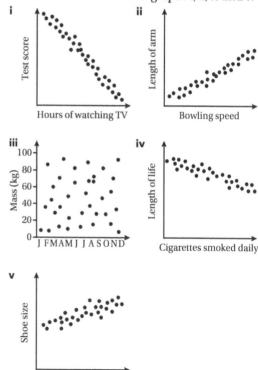

2 Sookie collected data from 15 students in her school athletics team. She wanted to see if there was a correlation between the height of the students and the distance they could jump in the long-jump event. She drew a scatter diagram to show the data.

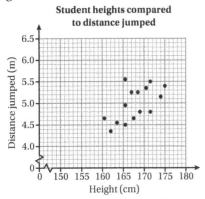

Student heights compared to distance jumped

a Copy the diagram and draw the line of best fit on to it.

b Use your line of best fit to estimate how far a student 165 cm tall could jump.

c For the age group of Sookie's school team, the girls' record for long jump is 6.07 m. How tall would you expect a girl to be who could equal the record jump?

d Describe the correlation shown on the graph.

e What does the correlation indicate about the relationship between height and how far you can jump in the long jump event?

3 Mrs Andrews wants to know whether her students' results on a mid-year test are a good indication of how well they will do in the GCSE examinations. The results from the test and the examination are given for a group of students.

Student	Mid-year mark	GCSE mark	Student	Mid-year mark	GCSE mark
Anna	78	73	Tina	92	86
Nick	57	51	Yemi	41	50
Sarah	30	39	Asma	75	64
Ahmed	74	80	Rita	84	77
Sanjita	74	74	Mike	55	58
Moeneeb	88	73	Karen	90	80
Kwezi	94	88	James	89	87
Pete	83	69	Priya	95	96
Idowu	70	63	Claudia	67	70
Sam	61	67	Noel	45	50
Emma	64	68	Wilma	70	64
Gibrine	49	54	Teshi	29	34

a Draw a scatter diagram with the GCSE results on the vertical axis.

b Comment on the strength of the correlation.

c Draw a line of best fit for this data.

d Estimate the GCSE results of a student who got 65 in the mid-year test.

e Comment on the likely accuracy of your estimate in part **d**.

HOMEWORK 38F

1 For the following sets of data, one of the three averages is not representative. In each case, state which average does not represent the data well and give a reason for your answer.

a 6, 2, 5, 1, 5, 7, 2, 3, 8

b 2, 0, 1, 3, 1, 6, 2, 9, 10, 3, 2, 2, 0

c 21, 29, 30, 14, 5, 16, 3, 24, 17

2 A scatter graph of the age and shoe size of 11 boys is shown below.

a Comment on the correlation.

b Identify any outliers.

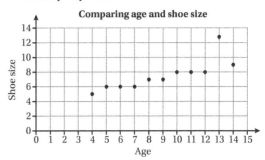

Comparing age and shoe size

3 Silvie works as a waitress. She records her tips for the last eight shifts.

£10   £20   £10   £15   £30   £25   £10   £200

a Find the mean amount she is tipped.

b What is the range of tips?

c What is the median amount she received in tips?

d What is the median without the outlier?

e What is the mean without the outlier?

4 Pete has collected 50 pieces of data about people's spending habits.
- The median spend is £27.85
- The mean spend is £34.70
- The minimum amount spent is £3.11 and the maximum is £93.34.

Would you expect there to be any outliers in this data set? Explain your answer.

Chapter 38 review

1 The mean of two consecutive numbers is 9.5. The mean of eight different numbers is 4.7.
  a Calculate the total of the first two numbers.
  b What are these two numbers?
  c Calculate the mean of the ten numbers together.

2 Three suppliers sell specialised remote controllers for access systems. A sample of 100 remote controllers is taken from each supplier and the working life of each controller is measured in weeks. The following table shows the mean time and range for each shop.

Which supplier would you recommend to someone who is looking to buy a remote controller? Why?

Supplier	Mean (weeks)	Range (weeks)
A	137	16
B	145	39
C	141	16

3 The ages of people who visited an art exhibition are recorded and organised in the grouped frequency table below.

Age in years (a)	Frequency
$0 < a < 10$	13
$10 < a < 20$	28
$20 < a < 30$	39
$30 < a < 40$	46
$40 < a < 50$	48
$50 < a < 60$	31
$60 < a < 70$	19
Total	

  a Estimate the mean age of people attending the exhibition.
  b What is the modal age group?
  c What is the median age of visitors to the exhibition?
  d Why can you not calculate an exact mean for this data set?

4 Study the scatter diagram and answer the questions.

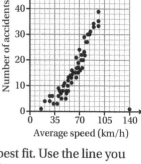

Accidents at a road junction

  a What does this diagram show?
  b What is the independent variable?
  c Place some tracing paper over the diagram and draw a line of best fit. Use the line you have drawn to predict:
    i the number of accidents at the junction when the average speed of vehicles is 100 km/h.
    ii what the average speed of vehicles is when there are fewer than ten accidents.
  d Describe the correlation.
  e What does your answer to **d** tell you about the relationship between speed and the number of accidents at a junction?
  f Comment on the outlier in this data set.

5 This cumulative frequency curve shows the masses of 500 12-year-old girls.

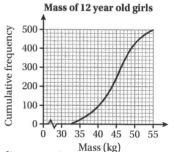

Mass of 12 year old girls

  a Estimate the quartiles for the data set.
  b What is the median mass?
  c Calculate the IQR for the data set.

6 The box plot shows the average distance (km) commuters in Leeds travel to get to work each day.

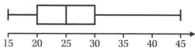

  a What is the median distance travelled?
  b What is the maximum distance travelled?
  c What percentage of commuters travel 30 km or less?
  d What percentage of the commuters travel between 15 and 25 km?
  e What is the IQR?
  f What is the range of distances travelled?

7 The box plot shows data for the times that people took to swim across the English Channel in 2013.

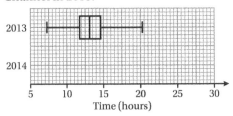

The data for swimmers in 2014 is as follows:

Shortest time = 7.1 hours
Lower quartile = 10.9 hours
Range = 18.5 hours
Median = 12.6 hours
Upper quartile = 14.8 hours

Compare the times taken by the swimmers in 2014 to the times in 2013.

39 Interpreting graphs

Section 1: Graphs of real-world contexts

HOMEWORK 39A

1 The graph below shows the path followed by a ball that is tossed into the air.

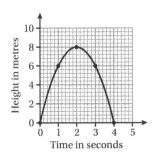

a What is the greatest height the ball reaches?
b How long did it take for the ball to reach this height?
c How high did the ball go in the first second?
d For how long was the ball in the air?
e Estimate for how long the ball was higher than 3 m above the ground.

2 Dabilo and Pam live 200 km apart from each other. They decide to meet up at a shopping centre in-between their homes one Saturday. Pam travels by bus and Dabilo catches a train. The graph shows both journeys.

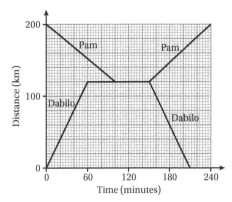

a How much time did Dabilo spend on the train?
b How much time did Pam spend on the bus?
c At what speed did the train travel for the first hour?
d How far was the shopping centre from:
   i Dabilo's home?
   ii Pam's home?
e What was the average speed of the bus from Pam's home to the shopping centre?
f How long did Dabilo have to wait before Pam arrived?
g How long did the two girls spend together?
h How much faster was Pam's journey on the way home?
i If they left home at 8:00 a.m., what time did each girl return home after the day's outing?

3 The population of bedbugs in New York City is found to have increased rapidly over a period of four months. The increases in number are given in the table.

Time (months)	Bedbug population (estimated)
0	1000
1	2000
2	4000
3	8000
4	16 000

a Plot a graph to show the increase over time.
b When did the number of bedbugs reach 10 000?
c Estimate the number of bedbugs there will be after six months if the population continues to grow at this rate.

Section 2: Gradients
HOMEWORK 39B

1 The graph shows the concentration of lactic acid in a runner's muscles before, during and after strenuous exercise.

a What is the normal amount of lactic acid in muscles (based on the graph)?
b What happens to the level of lactic acid after 10 minutes? How can you tell this?
c The runner stops exerting herself after 20 minutes. What happens to the levels of lactic acid after this?
d How long does it take the levels to return to normal?

2 This graph shows a cyclists journey from home to the post office.

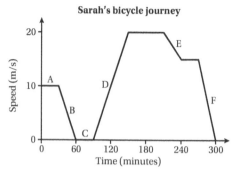

Match each label below to a labelled section of the graph.

stopped   constant speed   slowing down
quick decrease in speed to a stop
slowed to a stop   quick increase in speed

3 This graph shows what percentage of her salary Annabel spent paying bills.

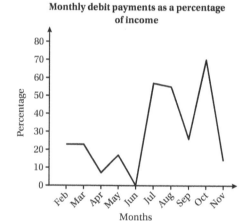

a What time period is represented on the graph?
b What percentage of Annabel's salary is used to pay bills in February?
c At what point does she have no bills to pay?
d When do her bill payments increase sharply? What might have caused this?
e One month Annabel uses almost all of her salary to repay a loan. When was this?
f Write a short description of Annabel's financial situation over the time period.

HOMEWORK 39C

1 The graph of $y = {}^-2x^2 + 4x + 5$ is shown below.

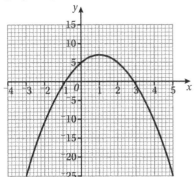

Find the gradient of the graph:
a where $x = 2$
b where $x = {}^-2$.

2 The graph represents a car journey.

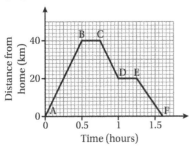

a What is the gradient when the graph is horizontal?
b What does this tell you about the car at these times?
c Use the gradient to determine the speed of the car for sections:
  i AB   **ii** CD   **iii** EF.

3 The graph shows how the growth rate of a bean plant is affected by light. Find the rate of growth with and without light on the thirteenth day.

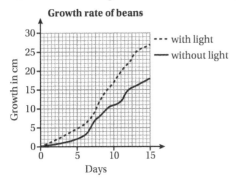

Section 3: Areas under graphs
HOMEWORK 39D

1 This speed-time graph shows the speed of a car in km/h against the time in minutes.

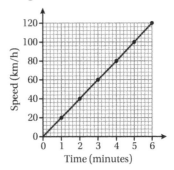

a What is the speed of the car after:
  i two minutes   **ii** six minutes?
b When is the car travelling at 70 km/h?
c Calculate the acceleration of the car in km/h².
d What distance did the car cover in the first six minutes?

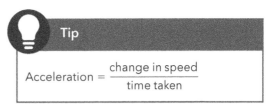

Tip

$$\text{Acceleration} = \frac{\text{change in speed}}{\text{time taken}}$$

2 This speed–time graph shows the speed of a train in m/s against the time in seconds.

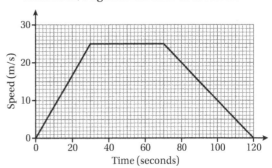

a When was the train accelerating and what was the acceleration?
b When did the train start decelerating and what was the deceleration?
c When the train was travelling at a constant speed, what was its speed in km/h?
d What distance did the train travel in two minutes?

3 This travel graph shows the journey of a petrol tanker doing deliveries.

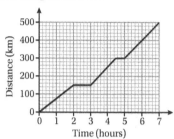

a What distance did the tanker travel in the first two hours?

b When did the tanker stop to make its first delivery? For how long did it stop?

c Calculate the average speed of the tanker between the first and second stop on the route.

d What was the average speed of the tanker during the last two hours of the journey?

e How far did the tanker travel on this journey?

4 This speed-time graph shows the speed in m/s for a car journey.

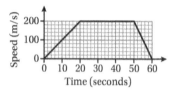

a For how long was the car accelerating?

b At what rate did it decelerate from 50 to 60 seconds?

c What distance did the car cover in the first 20 seconds?

d How many metres did the car take to stop once it started decelerating?

Chapter 39 review

1 This distance–time graph shows a cyclist's journey during a cross-country cycle race.

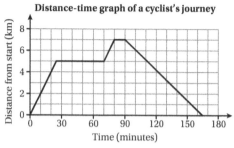

Distance-time graph of a cyclist's journey

a Calculate the cyclist's average speed for:
   i the first ten minutes of the race
   ii the whole race.

b How far was the cyclist from the start/finish point after two hours?

c The cyclist takes 45 minutes to fix a serious puncture. How far was the cyclist from the starting point when she got the puncture?

2 This graph shows the speed, in m/s, of a car as it comes to rest from a speed of 10 m/s.

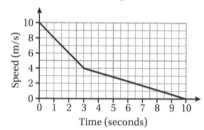

a Calculate the rate at which the car is slowing down during the first three seconds.

b Calculate the distance travelled during the 10-second period shown on the graph.

c Calculate the average speed of the car for this 10-second period.

40 Algebraic inequalities

Section 1: Expressing inequalities

HOMEWORK 40A

1 Write each of the following statements as an inequality.
List three integer values that satisfy each inequality.

a f is less than or equal to 4.
b x is more than 8 but less than 12.
c y is greater than 2 but less than 8.
d x is less than or equal to 12 but greater than 8.
e x is greater than 4 than but less than or equal to 9.

2 Write down three possible integer solutions for each of these inequalities. Give a value for x and y in each case.

a $x + y > 2$ **b** $x + y < 0$ **c** $x - y > 3$
d $x - y > 1$ **e** $xy \leqslant 6$ **f** $\frac{x}{y} \geqslant 4$

Section 2: Number lines and set notation

HOMEWORK 40B

1 Draw a number line to represent each inequality.

a $x > {}^-2$ **b** $x < {}^-3$ **c** $x \leqslant \frac{1}{2}$
d $x \geqslant 3$ **e** ${}^-1 < x < 3$ **f** $2 \leqslant x \leqslant 5$
g ${}^-3 \leqslant x < 0$ **h** ${}^-3 < x < 4$ **i** ${}^-2 < x \leqslant 4$
j ${}^-1 \leqslant x < 2$

2 Use set notation to describe the sets shown on each number line.

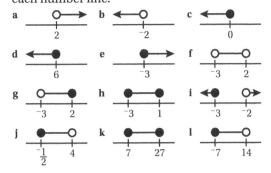

Section 3: Solving linear inequalities

HOMEWORK 40C

3 Draw a number line to represent each of the following.

a $\{x : x > 1\}$ **b** $\{x: {}^-1 < x \leqslant 1\}$
c $\{x: x < 0 \text{ or } x \geqslant 1\}$ **d** $\{x: {}^-2 \leqslant x < 2\}$

4 Is $\{x: x < {}^-1 \text{ or } x > 2\}$ equivalent to $\{x: 2 < x < {}^-1\}$? Explain your answer.

1 Solve for x.

a $3x \leqslant {}^-9$ **b** $2 - 4x > 1$ **c** $\frac{3x}{{}^-4} \geqslant {}^-6$
d $3 - 6x < {}^-8$ **e** $\frac{{}^-x}{4} < 8$ **f** $\frac{{}^-7x}{6} < {}^-7$

2 Solve each inequality. Show your solution on a number line.

a $1 - 2x > x - 2$ **b** $2(1 - x) < 5$
c $3(4 - x) > 12$ **d** $3x - 5 < x + 6$
e $4 < 2(2x - 3)$ **f** $\frac{2x - 1}{3} \leqslant 6$

3 Solve each of the following inequalities. Draw a number line to show each solution.

a $x \geqslant 4$ **b** $2x + 4 \leqslant 16$
c $6x + 12 \geqslant 48$ **d** $3(3x - 2) + 4 < 2(4x + 3)$
e $\frac{5x}{6} \geqslant {}^-5$ **f** $\frac{3x - 4}{3} \leqslant 17$
g $\frac{4(2x + 3)}{7} > 1$

4 Louise is x years old. Her sister Jayne is four years younger. The sum of their ages is less than 28. What are the possible ages that they could be?

5 This is the plan of an L-shaped exhibition space which must be at least 30 m² in area. What lengths must the sides marked x and $2x$ be to meet these conditions?

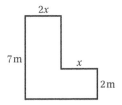

Section 4: Solving quadratic inequalities

HOMEWORK 40D

1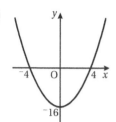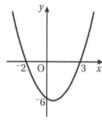

Use the related graphs to work out for what values of x:

a $x^2 > 16$ **b** $x + 6 \geqslant x^2$

2 Sketch a graph to solve each of the following inequalities.

a $x(x - 4) < 0$ **b** $(x + 1)(x - 3) \geqslant 0$
c $2x - x^2 \leqslant 0$

3 Solve for x. Show each solution on a number line.

a $(x - 5)(2x - 1) > 0$ **b** $(x + 3)(x + 1) \geqslant 0$
c $x^2 + 6x + 8 > 0$ **d** $4x \geqslant 3 + x^2 \leqslant 0$

Section 5: Graphing linear inequalities

HOMEWORK 40E

1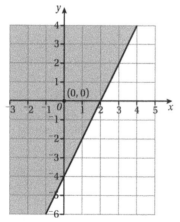

For this graph:

a Write the equation of the solid line.
b Write an inequality to describe the shaded area.
c Show algebraically that the point $(3, {}^-3)$ is not in the region defined by the graph.

2 Draw a graph to represent each inequality and shade the region represented by each.

a $y \leqslant 2x$ **b** $x + y \leqslant 4$
c $4x + 4y \leqslant 8$ **d** $y \geqslant x + 1$
e $4x + 2y > 12$ **f** $\dfrac{3x}{2} + 3y < 12$
g $\dfrac{1}{2x} + \dfrac{1}{3y} \leqslant \dfrac{1}{3}$ **h** $y \leqslant 4x$

HOMEWORK 40F

1 Graph the feasible region defined by $x \geqslant 0, y \geqslant 0, y + x \leqslant 10$ and $x + 2y \leqslant 16$.

2 Graphically solve the simultaneous linear inequalities:

a $x \geqslant 2$ and $y + 2x \geqslant 10$
b $y - x \leqslant 1$ and $y + 2x > 6$

3 Shade the region which satisfies the inequalities $x \geqslant 1, y \geqslant 2, y \leqslant 6 - x$ and $2x + 3y \leqslant 15$.

Chapter 40 review

1 Solve each inequality. Show the solution on a number line.

a $\dfrac{4x}{9} > 8$ **b** $2x + 7 < {}^-13$
c $2x - 17 \geqslant 15$ **d** $2(x + 5) > {}^-8$
e $4 - \dfrac{x}{4} \leqslant 7$ **f** $3(x - 2) \leqslant {}^-9$

2 Solve for x.

a $4x + 3 < 18$ **b** $10 + 3x \leqslant 9$
c $3(x + 2) > 12$ **d** $2(5x - 4) \geqslant 20$
e $y^2 > 9$ **f** $x^2 - x - 6 \leqslant 0$

3 Two sportsmen have played at least 44 games this season. One player is known to have played in 12 more games than the other. What is the least number of games each player could have played?

4 Draw a graph and shade the feasible region that satisfied the inequalities $x \geqslant 0, y \geqslant 0$, $10x + 5y \geqslant 100, x + y \geqslant 15$ and $2x + 4y \geqslant 40$.

41 Transformations of curves and their equations

Section 1: Quadratic functions and parabolas

HOMEWORK 41A

1. These sketch graphs show functions in the form of $y = ax^2 + q$.

a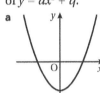

b

c

d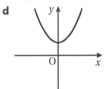

   a For each graph, state whether a and q are positive, negative or zero.

   b In the function, $y = ax^2 + q$, what is the effect on the graph if you decrease q by n?

2. A car accelerates from a stationary position at a traffic light for 10 seconds. The car's distance from the traffic light (in metres) after x seconds is given by the function

   $y = x^2 + 5$.

   a Represent this relationship graphically and explain the shape of the graph.

b Sketch the general parabola $y = x^2$ on the same grid.

c Extend your initial graph to include negative values of x.

What is the equation of the resultant graph?

d Compare the graphs of $y = x^2$ and $y = x^2 + 5$.

   i What features do the graphs have in common? Explain why.

   ii How are the graphs different? Explain why.

3. Given the equation $y = x^2 + q$, draw sketch graphs to show the effect on the graph when:

   a $q = 0$ **b** $q = 4$ **c** $q = \dfrac{1}{4}$

   d $q = {}^-4$ **e** $y = {}^-x^2 + 4$.

4. **a** Sketch the graphs of $y = {}^-x^2 - 2$ and the graph of $y = {}^-x^2$ on the same grid.

   b Compare the graphs.

   i What features do the graphs have in common? Explain why.

   ii How are the graphs different? Explain why.

HOMEWORK 41B

1. Complete the following statements for the graph of $y = ax^2 + q$.

   a If $a > 0$ and $q > 0$, the graph ...

   b If $a > 0$ and $q < 0$, the graph ...

   c If $a > 0$ and $q > 0$, the graph ...

   d If $a > 0$ and $q > 0$, the graph ...

2. **a** Sketch the graph of $y = \dfrac{{}^-1}{2x^2} - 1$ and the graph of $y = {}^-x^2$ on the same grid.

   b Compare the graphs.

   i What features do the graphs have in common? Explain why.

   ii How are the graphs different? Explain why.

   c Describe how increasing the value of a by n affects the graph.

   d How does decreasing the value of a by n affect the graph?

3 The graph $y = x^2$ is shown as a dotted line on grids **i** to **iv** for reference.

  a Complete the table for the other graph on each grid.

Equation	Coordinates of turning point	y-intercept
$y = (x + 3)^2$		
$y = (x - 1)^2$		
$y = (x - 2)^2$		
$y = (x + 4)^2$		

i $y = (x + 3)^2$

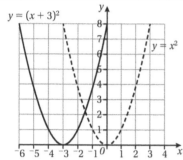

ii

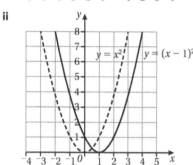

iii

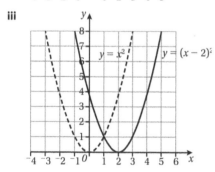

iv $y = (x + 4)^2$

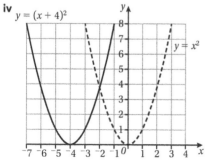

b Explain how the constant p in $y = (x + p)^2$ transforms the graph of $y = x^2$.

c Sketch the graph of $y = {}^-(x + 3)^2 + 6$ using what you know about shifting graphs.

4 The equation $y = {}^-(x + p)^2 + q$ is a transformation of the graph $y = x^2$ in which the graph is reflected in the x-axis and translated p units left and q units up.

Describe how the graph $y = x^2$ is transformed in each of the following cases:

  a $y = (x + p)^2 + q$ **b** $y = {}^-(x + p)^2 - q$

  c $y = (x - p)^2 + q$ **d** $y = {}^-(x - p)^2 - q$.

HOMEWORK 41C

1 Use what you know about transforming curves to sketch the following graphs.

  a $y = (x - 3)^2$ **b** $y = (x - 1)^2 - 2$

  c $y = x^2 - 8x + 13$

2 Factorise each quadratic equation and then sketch each graph.

  a $y = x^2 + 4x + 4$ **b** $y = x^2 - 8x + 16$

  c $y = x^2 + 16x + 64$

3 Complete the square on the equation $y = {}^-x^2 + 6x + 3$ and sketch the graph that results.

4   **a** Sketch the graph of $y = x^2 - 4x + 2$.

  b Explain how you could use this graph to sketch the graphs of:

    i $y = x^2 + 4x + 2$ **ii** $y = {}^-x^2 + 4x$

Section 2: Trigonometric functions

HOMEWORK 41D

1 Describe how the graph of $y = \sin x$ has been transformed to produce graphs A to D on the grid below and write the equation for each graph.

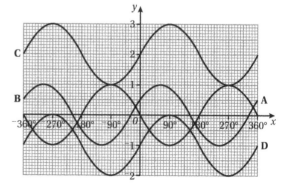

2 The graph below has the general equation
$y = \sin(x + a)$.

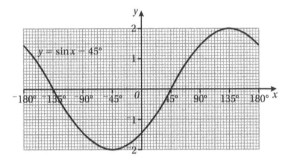

a What is the value of a?

b Over what period does this graph repeat?

c If this graph was shifted 30° left, what would the new equation of the graph be?

3 Explain why $\sin x = \cos(x - 90)$ is a trigonometrical identity.

4 Sketch the graph of $y = \cos(x + 45)$ and explain how it is different to the graph of $y = \cos x$.

5 Two graphs in the form of $y = \cos x + a$ have been drawn on the same grid.

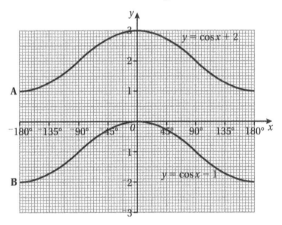

a Write the correct equation for each graph.

b How is transformation $y = \cos x + a$ different from the transformation $y = \cos(x + a)$?

6 Explain how the following transformations change the graph of $y = \tan x$.

a $y = {}^-\tan x$

b $y = \tan x + 2$

c $y = \tan x - 1$

Section 3: Other functions
HOMEWORK 41E

1 The graph of $y = x^3 - x$ is shown below.

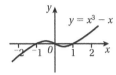

a Use this graph to sketch the graph of $y = x^3 - x + 4$.

b How is the graph of $y = x^3 - x$ transformed by adding 4 to the equation?

c How would the graph of $y = x^3 - x$ be transformed to produce $y = {}^-x^3 - x - 1$?

2 Complete these statements about translating curves.
Include the direction of the translation as well as the number of units.

a The graph of $y = x + 4$ is the graph of $y = x$ translated ...

b The graph of $y = x^3 - 1$ is the graph of $y = x^3$ translated ...

c The graph $y = \dfrac{1}{x}$ can be translated 2 units upwards to form the graph ...

d The graph of $y = (x + 2)^3$ translates the graph of $y = x^3$...

e The graph of $y = \dfrac{1}{x}$ can be translated 1 unit horizontally to the left to give the graph with the equation ...

3 The basic reciprocal graph $y = \dfrac{1}{x}$ has undergone a translation to produce graph B. What is the equation of graph B?

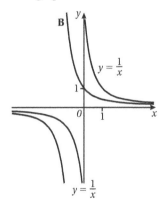

4 The graph of $y = 3^x$ is shown here.

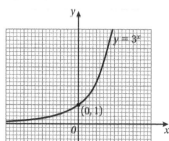

a Sketch the reflection of this graph in the y-axis and write its equation.

b Sketch the graph that results from translating $y = 3^x$ five units in the positive y-direction.

c How would the graph of $y = 5(3^x)$ differ from $y = 3^x$?

Section 4: Translation and reflection problems

HOMEWORK 41F

1 The graphs of parabolas **a** and **b** have been drawn on the same grid. Graph **b** is a translation of graph **a** $4\frac{1}{2}$ units downwards.

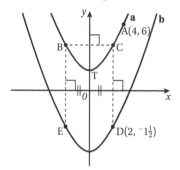

Determine:

a the equation of graph **a**

b the equation of graph **b**

c the coordinates of point B.

2 Fully describe the transformations which transform the graph of $y = (x + 1)^2$ to produce:

a $y = (x - 3)^2$

b $y = {}^-(x + 1)^2$

c $y = (x + 4)^2 - 2$

3 The graph of $y = {}^-x^2 + 4x$ is shown here.

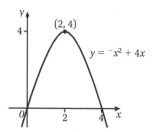

a Sketch the reflection of this graph in the y-axis and write the equation of the reflection.

b Reflect $y = {}^-x^2 + 4x$ in the x-axis. Write the equation of the reflected graph.

c What equation will give the reflection of the cubic graph $y = {}^-x^3 + 8$ in the y-axis?

d What can you conclude about the equation of a graph and its reflection in:

  i the y-axis?

  ii the x-axis?

4 The graphs A and B have been drawn on the same grid.

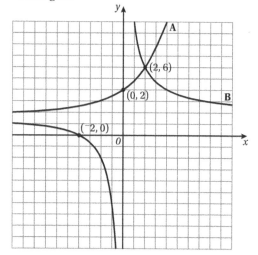

a Determine the equation of graph A.

b Determine the equation of graph B.

c Graph A is translated up two units. Give the new equation of the graph.

d The graph of B is reflected about the x-axis. What function will produce the new graph?

5 Two graphs of $y = \tan x$ have been transformed and drawn below.

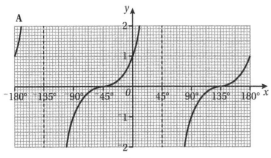

A

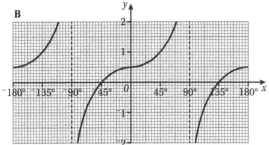

B

a What is the equation of graph A?
   What does this tell you about how the graph was transformed?
b What is the equation of graph B?
c How is this transformation different from the one in graph A?
d Where are the asymptotes of graph B?

Chapter 41 review

1 Copy and complete Table A at the bottom of the page to summarise what you have learned about transforming curves and their equations in this chapter.

2 Describe in words the transformations that have been applied to the original function to give the new equation in each case.
   a $y = x^2$ has been transformed to
      $y = (2x + 3)^2 - 5$
   b $y = \sin x$ has been transformed to
      $y = 3 \sin (x - 30) + 1$
   c $y = x^3$ has been transformed to $-2(x + 3)^3$
   d $y = 2^x$ has been transformed to $y = 2^{-x} +$
   e $y = \dfrac{1}{x}$ has been transformed to $y = \dfrac{1}{(x - 1)} + 4$

Table A

Graph and general form of equation	Horizontal translation	Vertical translation	Reflection in y-axis	Reflection in x-axis
Quadratic				
Reciprocal				
Exponential				
Cubic				
sin				
cos				
tan